AF802837

General Physics I

Physics Lab Workbook

S.H. Grzeszczuk

Felician University

Kendall Hunt
publishing company

All cover photos taken by S. H. Grzeszczuk. Photo collage creation © 2023 Gabriella LoPresti

Kendall Hunt
publishing company

www.kendallhunt.com
Send all inquiries to:
4050 Westmark Drive
Dubuque, IA 52004-1840

ISBN 979-8-3851-0090-3

Published in the United States of America

The author wishes to express her gratitude to
Dr. Robert Keesey,
whose questions and remarks
led to developing parts of the content of this publication.

A sincere thank you goes to
Prof. Ghassan Nazi,
whose passion for teaching physics and shared experience
has been a true inspiration.

The author addresses words of appreciation to
Dr. Michael Sanford,
whose readiness to discuss mathematical applications and attention to details
significantly improved the content of this workbook.

Contents

Laboratory	Date of Completion	Group Members
Lab 1. Statistical Analysis		
Lab 2. Measurement Techniques		
Lab 3. Air Track		
Lab 4. Kinematic Graphs		
Lab 5. Pendulum		
Lab 6. Projectile Motion		
Lab 7. Force Table		
Lab 8. Inclined Plane		
Lab 9. Friction Coefficient		
Lab 10. Collisions		
Lab 11. Torque and Rotations		
Lab 12. Buoyant Force		
Lab 13. Springs		
Lab 14. Simple Harmonic Motion		

Lab 1
Statistical Analysis

Introduction

The purpose of this lab is to investigate the sources of experimental error and apply statistical analysis to data collections.

Upon completion of this lab, students will

- Know the common sources of experimental error.
- Recognize random errors.
- Understand the sources of human error.
- Apply descriptive statistics to present experimental data.
- Present experimental results with statistical significance.

Part 1. Simple Measurement

In this part of the lab, we will measure the length of the lab table.

Step One. Identify the measurement tool that you will use in this experiment (a ruler, a yardstick, an app on your phone or tablet, any object of a known length—creativity is encouraged!)

Measurement tool and unit: ______________________________

Step Two. Measure the length of the table with your tool.
Important: DO NOT share the results with your teammates!

Step Three. Record the measurement in the following box.

x =

Now you can share the results with your colleagues.

Step Four. Collect the results obtained by your classmates in Table 1-1. Use this exercise to learn their names (Column 1).

Table 1-1 Length measurements.

#	Student Name	Measurement Value x_i	x_i [m]
	Column 1	Column 2	Column 3
1			
2			
3			
4			
5			
6			
7			
8			
9			
10			
11			
12			
13			
14			
15			
16			
17			
18			
19			
20			
21			
22			
23			
24			

Critical Evaluation of Dataset

Take a moment to look at the data (Column 2). Write down your spontaneous observations.

What units did your colleagues choose when presenting their measurements?

What is the proper International System of Units (SI) unit for length? ____________________

Use the third column of Table 1-1 to present the collected data in proper units.

Even if presented in the same units, are all measurements identical?

Why do the measurements vary? After all, you have measured the same length!

In your opinion, which measurement is correct? Why?

Are there any outliers in Column 3? (Outliers are values that significantly differ from other data; see the following Example.)

Example

Students were asked to measure the total length of the whiteboards mounted on a wall in their classroom. Here are the results:

Student Name	Measurement	Measurement in Standardized Units	L [m] (*)
Aasiyah	5.31 m	5.31 m	5.31
Brian	4 yards	3.6576 m	~~3.66~~
Chris	197 in.	5 m	5.00
Dakota	5.5 yd.	5.03 m	5.03
Ethan	16′ 8″	5.08 m	5.08
Fatima	17′	5.18 m	5.18
Giovanni	17 feet	5.18 m	5.18
Hoyam	5 m 20 cm	5.2 m	5.20
Ivana	50 yd.	45.72 m	~~45.72~~
Julia	5 meters	5 m	5.00

(*) In this table, two decimal places were used. Precision of measurements and significant figures will be discussed in the next lab.

The second column lists all measurements as recorded. You may notice that one cannot even evaluate the collected data based on the second column due to the various units used. Thus, all measurements are standardized by converting them to meters in the third column. Finally, the fourth column presents only the numerical values, with the appropriate unit stated in brackets in the head row.

A first look at the data presented in the last column reveals that about 5*m* is the most frequent value. There are two significantly different values, called outliers; these are 3.66*m* and 45.72*m*. We may ask why those two values are significantly different. Presumably, Brian "lost" one yard somewhere in the process of taking the measurement. The 50 yards reported by Ivana looks like a typo ("50" instead of "5"). These are examples of ***human error***. Regardless of the reasons for human error, we **exclude** such **outliers** from further analysis. Including values that carry significant error may skew the results.

> ***Remember:*** *Never reject data rashly. Excluding outliers should be carefully considered and explained in the lab report along with the rationale behind the decision.*

Part 2. Estimate of the Measured Quantity and Error Analysis

No single measurement recorded in Table 1-1 represents the true value. The arithmetic mean, or the average of the data, is the best estimate of the measured value. The average is defined as

$$\bar{x} = \frac{1}{n}\sum_{i=1}^{n} x_i$$

where x_i is a singular measurement, and n is the total number of measurements. The Greek letter Σ [sigma] indicates the sum of all measurements x_i, from $i = 1$ (written below Σ) to n (written above Σ).

Based on the data collected in Table 1-1, the estimation of the value measured in the lab today is

$\bar{x} =$ ______________________________

Statistical analyses are further used to describe the variance of the dataset and the error of the mean. Sample variance s^2 is a number that describes the spread of measured values around the calculated average, and is defined as follows:

$$s^2 = \frac{1}{n-1}\sum_{i=1}^{n} (x_i - \bar{x})^2$$

The square root of the sample variance is called the standard deviation of the sample. Dividing the sample standard deviation by the square root of the total number of measurements n leads to the standard error $SE_{\bar{x}}$:

$$SE_{\bar{x}} = \frac{s}{\sqrt{n}}$$

The SE of the mean of data collected in Table 1-1 is

$SE_{\bar{x}} =$ ______________________________

Wait a minute...
Do we calculate all of that by hand?

Many statistical values can be calculated in Excel; see the following continuation of the Example.

Example (cont.)

Many statistics can be calculated in Excel. Formulas for mean (average), sample variance, and standard deviation are shown in Figure 1-1, cells D2, D4, and D5, respectively. Standard error is found by dividing the standard deviation (cell E5) by the square root of the number of recorded measurements (cell E6). In this case, we have just eight measurements, which makes it easy to count or remember. If we have more measurements, Excel offers a counting function shown in cell D6 of Figure 1-1. Finally, standard error of the mean is calculated as presented in cell E7 of Figure 1-1. Data collected by students in our example (without outliers) is presented in cells A1:A8. Respective numerical values are recorded in Column E.

	A	B	C	D	E
1	5.31				
2	5.00		*mean*	=AVERAGE(A1:A8)	5.1225
3	5.03				
4	5.08		*sample variance*	=VAR.S(A1:A8)	0.012592857143
5	5.18		*sample standard deviation*	=STDEV.S(A1:A8)	0.112217900278
6	5.18		*sample size n*	=COUNT(A1:A8)	8
7	5.20		*standard error*	=E5/SQRT(E6)	0.039675019129
8	5.00				

Figure 1-1 Screenshot of data analysis in Excel.

In Figure 1-1, the statistical measures are presented for learning purposes. All calculations can be done in one cell by using the following cell code:

$$= STDEV.S(A1:A8)/SQRT(COUNT(A1:A8))$$

where A1:A8 are cells that contain data.

In our example, the best estimation of the true length of the board is $5.1225m$ (cell E2). Standard error of the mean, as calculated in Excel, is 0.039675019129 (cell E7).

Standard error of the mean is used to report the experimental value of the measured quantity:

$$(\bar{x} \pm SE_{\bar{x}})\ unit$$

In this example, the measured length L is presented as

$$L = (5.1225 \pm 0.0397)\ m$$

Notice that both values were presented with only four decimal places (the least number of decimal places in the two calculated values). In fact, when analyzing further, we find that the fourth decimal place in the presented number represents tenths of a millimeter and is not significant for the length of the whiteboard. Thus, the results can be presented as

$$L = (5.123 \pm 0.040)\ m$$

Rules for deciding on how many significant figures or decimal places should be used will be discussed in the next lab. You will learn that sometimes percent uncertainty is needed. Percent uncertainty is calculated as

$$\frac{SE}{\overline{x}} \cdot 100\ \%$$

and, in the previous example, it is

$$\frac{0.040}{5.123} \cdot 100\% \cong 0.8\%$$

Percent uncertainty is a measure of reliability of the results. In general, the smaller the percent uncertainty, the better. In this course, some measurements might be accepted with 1% uncertainty, whereas others will be accepted with 5% uncertainty. Large percent uncertainty or percent error requires extensive error analysis.

Part 3. Presentation of the Results

Based on the earlier calculation of $\overline{x}$ and SE, the results of today's measurement exercise in the lab are:

$$L = (\_\_\_\_\_\_\_\_\_\_ \pm \_\_\_\_\_\_\_\_\_\_)\ m$$

Congratulations, you just presented
your first measurement in the physics lab!

Critical Thinking

FAQ: How many measurements should we plan?

Answer: Somebody said that the more data gathered, the better the estimate of the true value. However, that may be only partially correct. Consider the graph in Figure 1-2. When the number of measurements increases, standard error of the mean decreases proportionally to $\frac{1}{\sqrt{n}}$. Notice that adding 4 more measurements to the 6 already collected significantly decreases the error. Thus, 10 measurements significantly improve the results as compared to 6 measurements. However, for more than 16 measurements, the relative decrease by the $\frac{1}{\sqrt{n}}$ factor is less than 3%. For more than 50 measurements, adding 10 extra measurements will not improve the results more than 1%. So, plan enough measurements to minimize the percent uncertainty, but remember that the plotted points asymptotically approach the horizontal axis, so multiplying measurements does not promise obtaining the true value.

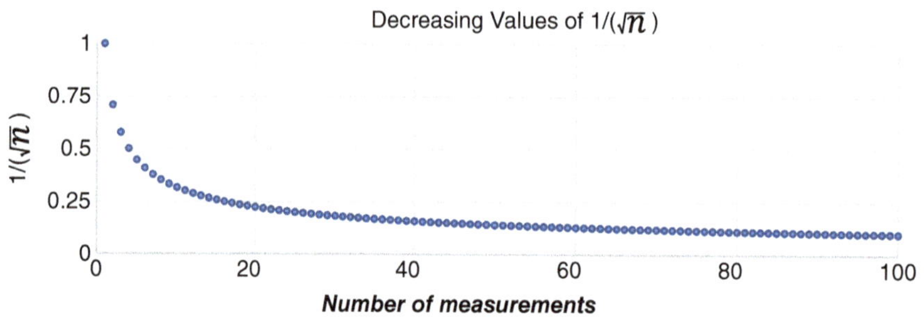

Figure 1-2 Discrete values of $\frac{1}{\sqrt{n}}$ as a function of increasing number of measurements.

Five-Minute Reflection

Can we ever know the "true values" of measured quantities?

Discuss in your group.

Takeaways

Use the following words to complete the following sentences:

average	reliable	units	reliability
outliers	uncertainty	random	carefully

1. It is up to the researcher to decide how many measurements should be taken to obtain ____________________ results. Reasoning behind the decision, supported by calculations, should be presented in the lab report.

2. Single measurements carry ________________________ error.

3. Measurements should be taken in proper _________________.

4. ____________________ should be excluded from data analysis because they carry significant errors and may skew the results. Excluding outliers should be ____________________ considered and explained in the lab report along with the rationale behind the decision.

5. _____________________ is the best estimate of the measured value.

6. Experimental results should be presented along with _____________________.

7. Percent uncertainty is a measure of _________________ of the results.

Lab 2

Measurement Techniques

Introduction

The purpose of this lab is to learn and apply various measuring techniques and evaluate these methods by comparing the results.

Upon completion of this lab, students will

- Know the three basic methods of measuring volume.
- Understand the limitations of measuring instruments.
- Estimate the measurement error based on the measurement instrument.
- Apply the rules of error progression.

Theory

1. What is volume? Explain it in your own words. If you cite a definition of volume, clearly indicate the source.

2. There are a few lab methods available to find the volume of an object. One of them is based on **mathematical formulas**.

   Write down the formulas for calculating the volume of the 3D shapes printed in Figure 2-1.

   $V_{rectangular\ prism} =$ ____________ $V_{sphere} =$ ____________ $V_{cylinder} =$ ____________

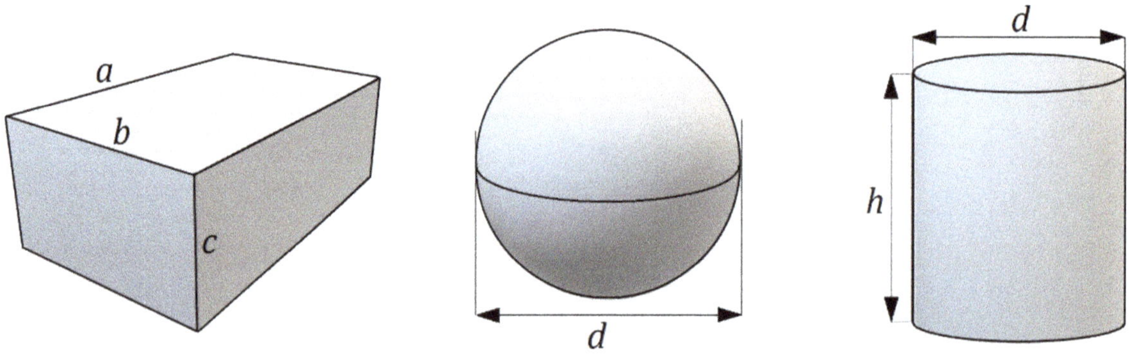

Figure 2-1 Rectangular prism, sphere, and cylinder.

3. Calculating the measurement error can be tricky with the formula method.

Critical Thinking

Let the width, length, and height of the rectangular block shown in Figure 2-1 be 2.0 ± 0.1 *cm*, 5.5 ± 0.1 *cm*, and 1.65 ± 0.03 *cm*, respectively. The product of width, length, and height of the block is 18.15 (check on your calculator). If the measurement errors are multiplied similarly to the dimensions, the error is 0.0003 (once again, check on your calculator). That would lead to a false conclusion that the volume is measured more accurately than any of its dimensions.

Thoughts?

Error Progression Theory

Apply the error progression theory to estimate the error for a volume of the 3D shapes if the volume is calculated with the previously listed formulas.

4. The previously described method of measuring the volume is used for objects of regular geometric shapes. To measure the volume of objects of irregular shapes, the water displacement method is more suitable. Based on Figure 2-2, explain to your team the water displacement method.

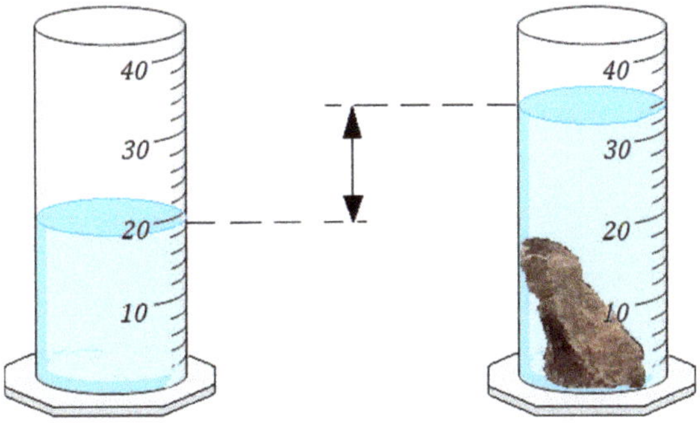

Figure 2-2 Water displacement method.

5. When reading the water level, one should maintain the proper line of sight as shown in Figure 2-3.

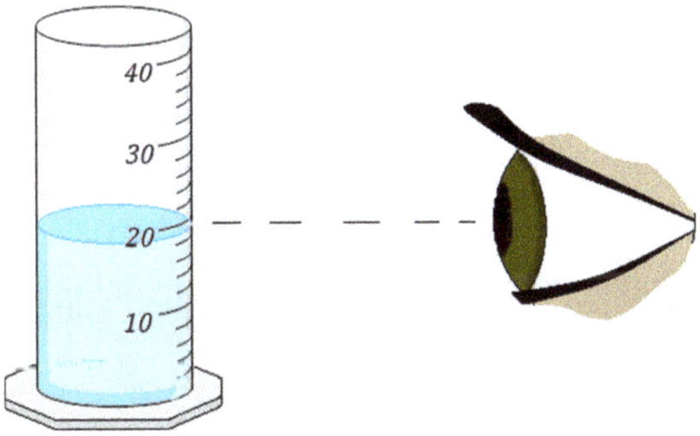

Figure 2-3 Proper water level reading.

Find at least two reasons why ***not*** maintaining the proper line of sight when reading the water level might cause an error.

Reason 1:

Reason 2:

6. The accuracy of the measurement is usually estimated as half of the smallest subdivision of the measurement instrument. For example, the smallest subdivision of the water cylinder shown in Figure 2-4 is $10\ mL$; thus, the accuracy of the measurement is $\pm 5\ mL$.

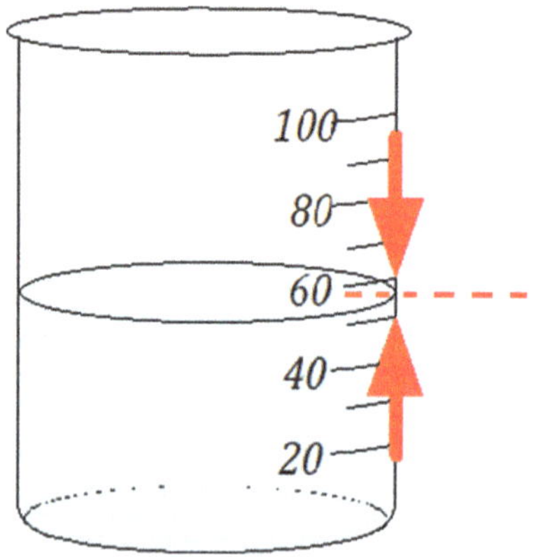

Figure 2-4 The smallest subdivision printed on a glass beaker.

Depending on the instrument, the smallest subdivision might be difficult to read.

The subdivision of the ruler shown in Figure 2-5 is $1\ mm$ and it is difficult to see a fraction of it. Hence, in this case, the accuracy of the measurement is $\pm 1\ mm$.

Figure 2-5 The smallest subdivision on a ruler.

Takeaway

The researcher must estimate the accuracy of performed measurements. When doing so, consider the instrument, the method of measurement, and the human condition of the researcher. If in doubt, use the method with the least accuracy (the greatest error) associated with the measurement and provide an explanation in the lab report.

Exercise

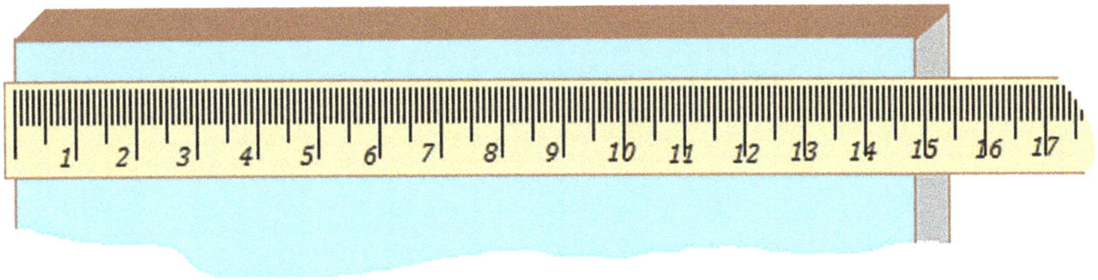

Figure 2-6 Measuring the length of a box.

What is the length of the box shown in Figure 2-6?

L = (________ ± ________) ________

7. The accuracy of digital instruments is provided by the manufacturer along with the range to which it applies.

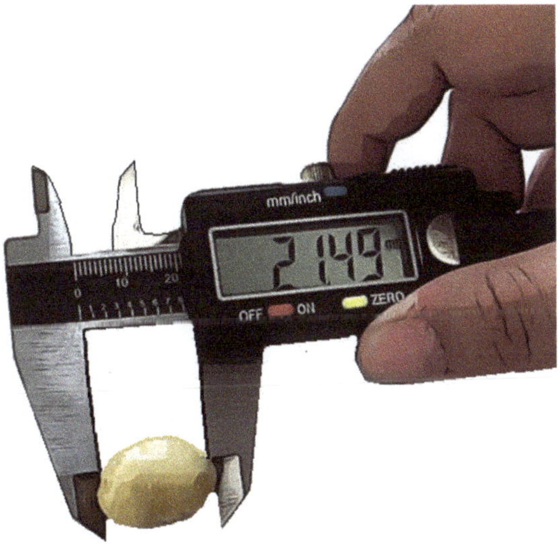

Figure 2-7 Measuring the length of a seed with an electronic caliper.

According to the manual for the digital caliper presented in Figure 2-7, the resolution of the instrument is $0.01\ mm$, and the accuracy is $\pm 0.02\ mm$. Thus, the length of the seed shown in Figure 2-7 is

$$L = (21.49 \pm 0.02)\ \text{mm}$$

8. Another way to calculate the volume is based on the definition of density. If density is the ratio of mass and volume, the volume can be calculated by dividing mass by density.

$$V = \frac{m}{\rho}$$

Exercise

Find the density of the following metals:

a. Aluminum ____________________
b. Bronze ____________________
c. Copper ____________________
d. Lead ____________________
e. Nickel ____________________
f. Steel ____________________
g. Zinc ____________________

9. When using a lab scale to find the mass of an object, remember to check for the measurement's accuracy. You can check the manual provided by the manufacturer. If the manual is not available, perform a simple assessment of the measurement's accuracy.

Lab Scale Accuracy Assessment

Step One. Turn on the scale and zero it (the scale should show a zero when nothing is placed on it).

Step Two. Weigh a known mass, such as 0.1 g, on the lab scale. Write down all the numbers shown on the scale.

Step Three. Turn off the scale.

Repeat Steps One to Three a few times. The variance of the readings will give you a sense of accuracy of a single measurement. Remember that, depending on the instrument, precision and accuracy are not necessarily the same.

Part 1. Volume of a Rectangular Block

In this part of the lab, you will measure the volume of a rectangular block. Please keep in mind that the volume of the block is not as important as the skills you will acquire.

Step One. Select a proper instrument to measure the length, width, and height of the rectangular block.

Instrument: ______________________________

Accuracy: ______________________________

Step Two. Measure the dimensions of a block. Repeat the measurement at least six times. Collect the data in Table 2-1.

Table 2-1 Measuring the dimensions of a rectangular block.

Measurement	Height (cm)	Length (cm)	Width (cm)
1			
2			
3			
4			
5			
6			
Average			

Step Three. Calculate the average of your measurements. Use the averages to calculate the volume of the block.

Step Four. Find the error of the volume. Use the error progression theory that you described at the beginning of this lab exercise.

Step Five. Express the results of your measurement and indicate the accuracy. Remember the units.

$$V_{\text{block}} = (\_\_\_\_\_\_\_ \pm \_\_\_\_\_\_\_)\ \_\_\_\_\_\_\_$$

Part 2. Volume of a Marble

In this part of the lab, you will measure the volume of a marble using two different methods: the water displacement method and the formula method. Then, you will calculate the percentage difference between the results to compare the methods. Please keep in mind that the volume of the marble is not as important as the skills you will acquire in this lab.

Water Displacement Method

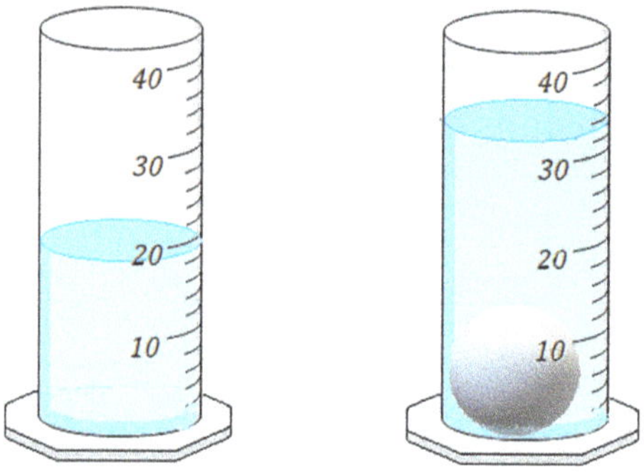

Figure 2-8 Measuring the volume of a sphere (water displacement method).

Step One. Fill the cylinder with water (Figure 2-8). Read the water level and write it down in the following line of text.

Water level before submerging the marble = ________________

Step Two. Gently put the object into the cylinder. Read the water level (Figure 2-8) and write down the measurement in the following line of text.

Water level after submerging the marble = ________________

Step Three. Estimate the accuracy of the measurement.

Present the results:

V_1 = (__________ ± __________) ____________

Formula Method

Step One. Use the caliper to measure the diameter of the marble.

Remember that the diameter is the largest dimension of the sphere. The most common mistake is to place the caliper jaws slightly off-center of the sphere (Figure 2-9a). To measure the diameter correctly, make sure that the sphere passes through the caliper as shown in Figure 2-9b. If you do not have a caliper, use the triangle method presented in Figure 2-9c.

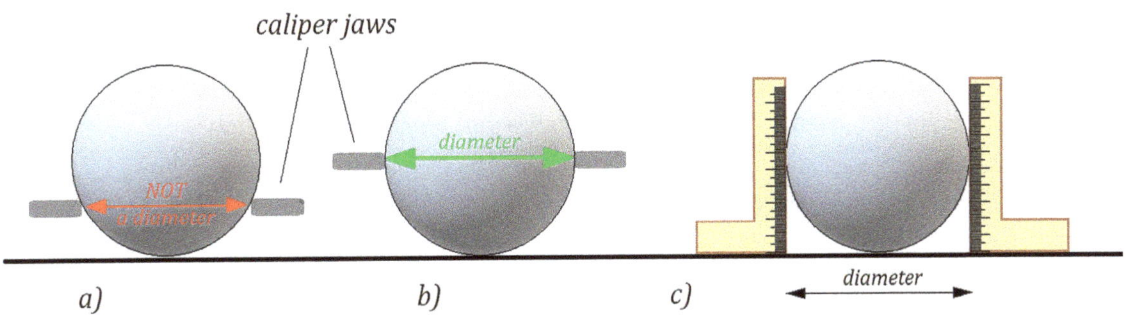

Figure 2-9 Measuring the diameter technique.

Step Two. Find out the accuracy of the measurement instrument. Present the results in the following line of text.

$$d = (\_\_\_\_\_\_\_\_ \pm \_\_\_\_\_\_\_\_)\ \_\_\_\_\_\_\_\_$$

Step Three. Use the formula to calculate the volume of the marble.

Step Four. Use the error progression theory to estimate the accuracy of the calculated volume of the sphere.

Step Five. Present the results:

$$V_2 = (\_\_\_\_\_\_\_\_ \pm \_\_\_\_\_\_\_\_)\ \_\_\_\_\_\_\_\_$$

Comparing the Results

Before comparing the results, you must present both in the same unit. Remember, that

$$1\ cm^3 = 1\ mL$$

Find the percentage difference between the results obtained with the water displacement method and the formula method.

$$\frac{\left|V_{1st\ method} - V_{2nd\ method}\right|}{average\ (V_1, V_2)} \cdot 100\%$$

Critical Evaluation of the Results

1. Are the results obtained with the two methods similar?

2. Which method is more accurate? Why?

Part 3. Volume of a Wire

In this part of the lab, you will measure the volume of a wire using two different methods of your choice. Then, you will calculate the percentage difference between the results to compare the methods. Be prepared to justify your choices of each method. There is no right or wrong method; it is the reasoning behind choosing the method that matters.

The First Method

Describe the first method.

Estimate the accuracy of the measurement: ______________________

Present the results.

$$V_1 = (\_\_\_\_\_\_\_\_ \pm \_\_\_\_\_\_\_\_)\ \_\_\_\_\_\_\_\_$$

The Second Method

Describe the second method.

Estimate the accuracy or error of the measurement: ______________________________

Present the results.

$V_2 = ($ __________ $\pm$ __________ $)$ __________

Comparing the Results

Find the percentage difference between the results obtained with the two methods.

$$\frac{|V_{1st\ method} - V_{2nd\ method}|}{average\ (V_1, V_2)} \cdot 100\%$$

Critical Evaluation of the Results

1. Are the results similar?

2. Which method is more accurate? Why?

Reflection

Fill in the blanks.

This was a ____________________ and ________________ lab. I worked with ____. New skills that I practiced today, such as ______________________________ and ______________________, will be useful in my future job. The main takeaway is ____.

Lab 3
Air Track

Introduction

The purpose of this lab is to investigate and graph the rate of change of distance in uniform and accelerated motion.

Upon completion of this lab, students will

- Describe distance in uniform motion as a function of time.
- Describe distance in accelerated motion as a function of time.
- Create a graph and adjust its details in Excel.
- Present experimental results on a graph that meets all standards of scientific graphing.

One-Minute Mindfulness Exercise

Before the class begins, stop for a quick minute. Take a deep breath and decide to commit your time to learning. You will not benefit from this lab if your mind is elsewhere, such as arguing with somebody you met in the morning or planning how to spend your evening.

This class is a part of your first course of physics. Nobody expects you to know all the answers and demonstrate skills. It is all right to ask and to not know.

You are learning kinematics. You are familiar with the nature of physics, methods of taking measurements, presenting results in proper units, and estimating measurement errors.

Put your cell phone down. You will connect back right after the lab.

Chat with your teammates before the lab begins. A good atmosphere in the group enhances learning.

Personal Note

Before this lab, I feel... ☹ 😐 😊

Because...

Theory

Before you come to the lab, make sure that you can answer the following questions. Please refer to your textbook to find the answers.

1. What is velocity?

2. What is acceleration?

3. What is the difference between instantaneous and average velocity?

4. What function is used to model distance as a function of time in uniform motion?

5. Write a formula for the $d(t)$ function, when $\vec{v} = const$.

6. What function is used to model distance as a function of time in accelerated motion?

7. Write a formula for the $d(t)$ function, when $\vec{a} = const.$

Part 1. Distance in Uniform Motion

You will use an air track to observe motion with no friction. Here are the step-by-step instructions on how to set the air track.

Step One. Experimental settings.

Level the air track before you start measurements. To do so, launch the air track engine and place the glider on. Observe the glider. If it is moving, the track is not leveled. Use the adjustable leg to level the track. Be precise and patient. This step is very important; the results of the experiment depend on how well you level the track.

Set the air track and photogates as shown in Figure 3-1.

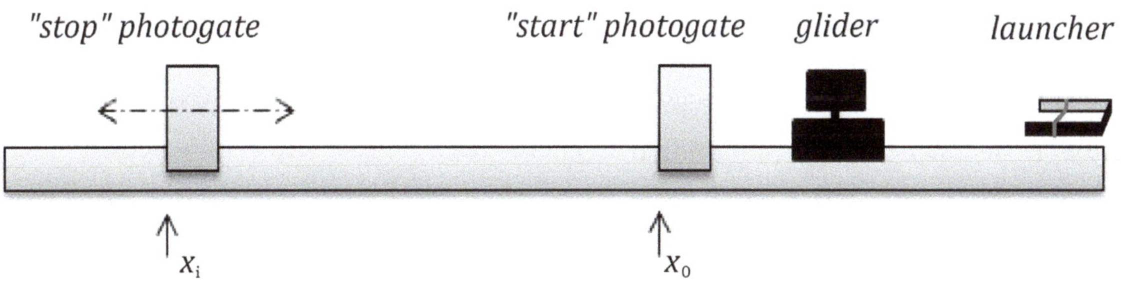

Figure 3-1 Distance in uniform motion experimental settings.

Mount the photogates on the air track. Place the first photogate ("start") in a position that will allow you to use the launcher. The "start" photogate will be in the same position, x_0, during the experiment. Place the second photogate (the "stop" one) close to the first one. You will be moving the second photogate along the air track to obtain measurements at various distances. Make sure that the photogates are mounted high enough to allow the glider to pass smoothly under both.

Install the rubber band launcher. The launcher's purpose is to keep the glider's speed the same (within the limits of error) for all measurements.

Connect the photogates to the electronic stop clock.

Step Two. Trial run.

Turn on the stop clock. It should show 00.0000 seconds. If it does not, then try the restart button. The photogates measure the time that the glider needs to travel the distance between them. When the glider passes through the first photogate, the clock starts measuring time; it stops when the glider passes through the second one. If the photogates operate in a different mode, try the MODE button.

When the photogates work correctly, you are ready to collect data.

Step Three. Data collection.

Place the first photogate in a position that will allow you to launch the glider. This photogate will be in the same position, x_0, during the experiment.

$$x_0 = \_\_\_\_\_\_\_\_\_\_\_\_\_\_\_\_\_\_\_\_$$

Set up the second photogate as close to the first one as possible. Place the glider in front of the first photogate, gently push it against the launcher, turn the air source on, and release the glider. Record the time of motion between the two photogates. Note the second photogate position x_1.

Move the second photogate about 5 centimeters farther from its original position and note x_2. Launch the glider and record the time of motion in Table 3-1.

Collect at least 20 data points. Record your data in Table 3-1.

Table 3-1 Distance in uniform motion data collection.

i	x_i	Time t [s]	i	x_i	Time t [s]
1			11		
2			12		
3			13		
4			14		
5			15		
6			16		
7			17		
8			18		
9			19		
10			20		

Teamwork

Make sure that all students in your team know what is going on and everybody has a task assigned. Possible tasks include turning the air source on and off and resetting the first photogate, launching the glider, bringing it back to the start position, moving the second photogate, and taking notes of the time. After a few measurements, switch roles.

Quick Evaluation

Congratulations, you are halfway through!

	Strongly agree	Agree	Neutral	Disagree	Strongly disagree
I am learning something.	o	o	o	o	o
The data collection is going well.	o	o	o	o	o
I am doing my best.	o	o	o	o	o
I contribute to the group work.	o	o	o	o	o
I have a feeling of accomplishment	o	o	o	o	o

Part 2. Distance in Accelerated Motion

In this part of the lab, you need to slightly adjust the experimental settings to use gravitational acceleration to move the glider. Elevate one side of the air track by placing a flat object under the right leg of the air track (see Figure 3-2). You can remove the rubber band launcher from the air track. The glider will accelerate due to gravity.

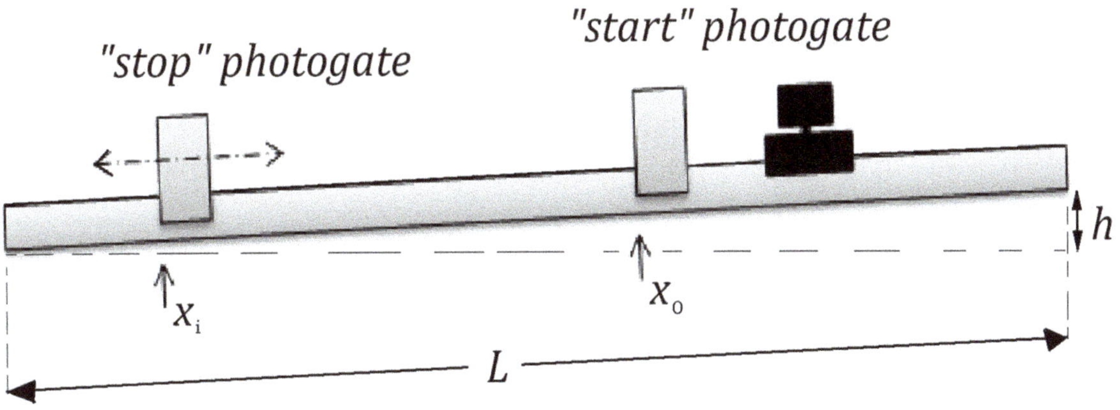

Figure 3-2 Distance in accelerated motion experimental settings.

Step Four. Measure the angle of elevation θ of the air track. Talk in your group and find the best method to measure the elevation angle. Describe the measurement method (you can use words or sketch a diagram which shows the method).

$\theta =$ ____________________

Step Five. Data collection.

Repeat the data collection similarly as you did in Step Three. Use Table 3-2 to record the second photogate position and the time of motion.

$x_0 =$ ____________________

Table 3-2 Distance in accelerated motion data collection.

#	x_i	Time t [s]	#	x_i	Time t [s]
1			11		
2			12		
3			13		
4			14		
5			15		
6			16		
7			17		
8			18		
9			19		
10			20		

Data Analysis

Begin with copying the collected data to Excel. We want to see distance as a function of time, so time is in the first column as shown in Figure 3-3.

	A	B	C	D	E
1	t[s]	x_i [cm]	d [m]		x_0 = 12 cm
2	0.2485	32			
3	0.3714	42			
4	0.4747	48			
5	0.5518	56			
6	0.6484	62			
7	0.7235	71			
8	0.8442	82			

Figure 3-3 Data columns in Excel.

Remember that your data collection includes positions of the second photogate on the air track, not the distances traveled by the glider. Thus, you need to calculate the distance. Also, you need to convert from centimeters to meters by dividing all results by 100. Excel can do both operations in a split second; just program cell C2 as follows:

$$= (B2 - x_0)/100$$

In the sample dataset used here, type in cell C2:

$$= (B2 - 12)/100$$

and copy it to all the cells in Figure 3-4.

COUNTIF fx = (B2-12)/100

	A	B	C	D	E
1	t[s]	x_i [cm]	d [m]		x_0 = 12 cm
2	0.2485	32	= (B2-12)/100		
3	0.3714	42	0.30		
4	0.4747	48	0.36		

Figure 3-4 Programming and copying cells.

To plot the data points, go to "Insert", choose "Charts", then choose "Scatter" with no lines or dot-to-dot connections.

Type the title of your graph and label the axes. The title should specify what is represented in the graph. Axes labels should indicate units.

By default, the markers are dots. Sometimes, however, dots as markers are not precise enough. Find "Marker Options" in the menu and adjust the markers (change the shape, size, and color).

Add error bars. By default, Excel will show both vertical and horizontal error bars. Since the photogates are quite precise, you can delete horizontal error bars. Choose "Standard Error" from the available options.

Add the trendline. In science, we would rather talk about the best fit line; in Excel, it is named a trendline.

Now, your graph should look like the one presented in Figure 3-5.

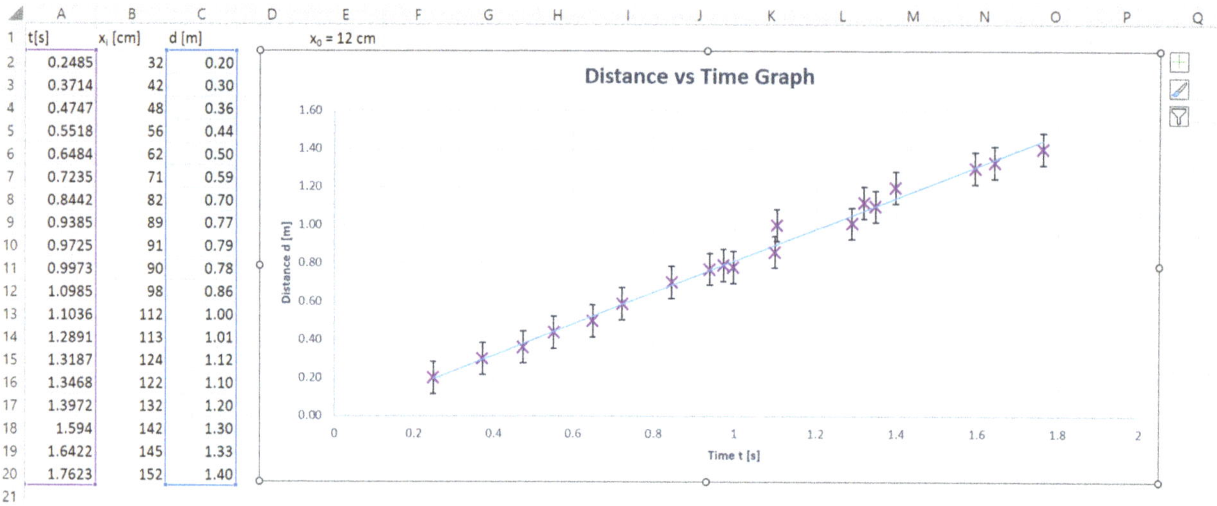

t[s]	x_i [cm]	d [m]
0.2485	32	0.20
0.3714	42	0.30
0.4747	48	0.36
0.5518	56	0.44
0.6484	62	0.50
0.7235	71	0.59
0.8442	82	0.70
0.9385	89	0.77
0.9725	91	0.79
0.9973	90	0.78
1.0985	98	0.86
1.1036	112	1.00
1.2891	113	1.01
1.3187	124	1.12
1.3468	122	1.10
1.3972	132	1.20
1.594	142	1.30
1.6422	145	1.33
1.7623	152	1.40

Figure 3-5 Sample graph, with distance as a function of time in uniform motion.

Check the proper box to see the equation of the line.

Plot the data collected in Table 3-2 in Excel. All steps are the same as in the earlier data analysis except for the best fit line (the trendline). According to the theory, the data points should form a parabola. To graph a parabola in Excel, go to the trendline options and choose polynomial of a second order (Figure 3-6). Your graph should look like the one shown in Figure 3-7.

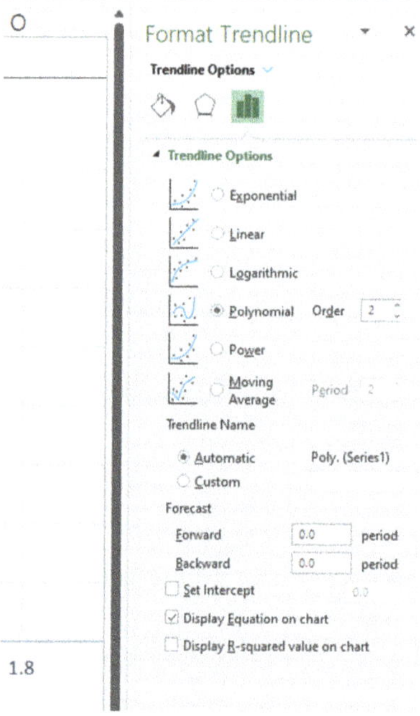

Figure 3-6 Graphing a parabola in Excel.

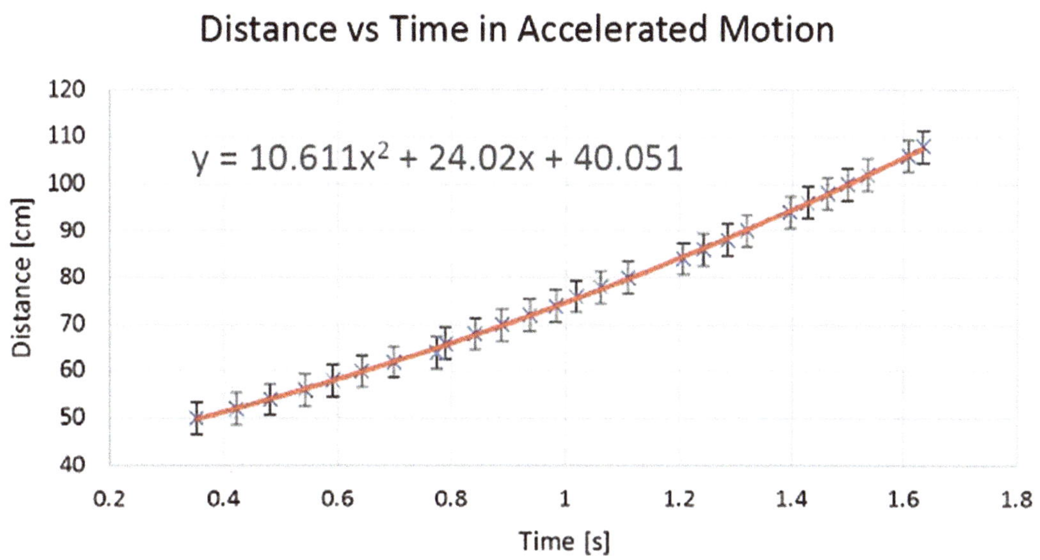

Figure 3-7 Sample graph, with distance versus time in accelerated motion.

Critical Thinking

The graph in Figure 3-7 is good but not perfect. Can you find the mistakes? Talk to your team and circle the mistakes in Figure 3-7.

FAQ–Graphing

Q: Do I have to put the data in order from the smallest to the largest?

A: That question is asked when students place the second photogate randomly at the air track. The answer is, not really. Data points will appear on the graph "in order". However, as you attempt to do research, you may develop a habit of taking measurements in order. It does not have to be the same order all the time, but some logic behind the data collection method shows your research culture. After all, research is systematic, not random, data collection.

Q: All data points are in one side or corner of the graph. What should I do?

A: This is a good question. Space management is important if you want to present your data. The first possible problem that comes to mind is that the axes are not adjusted to the range of plotted points (see Figure 3-8).

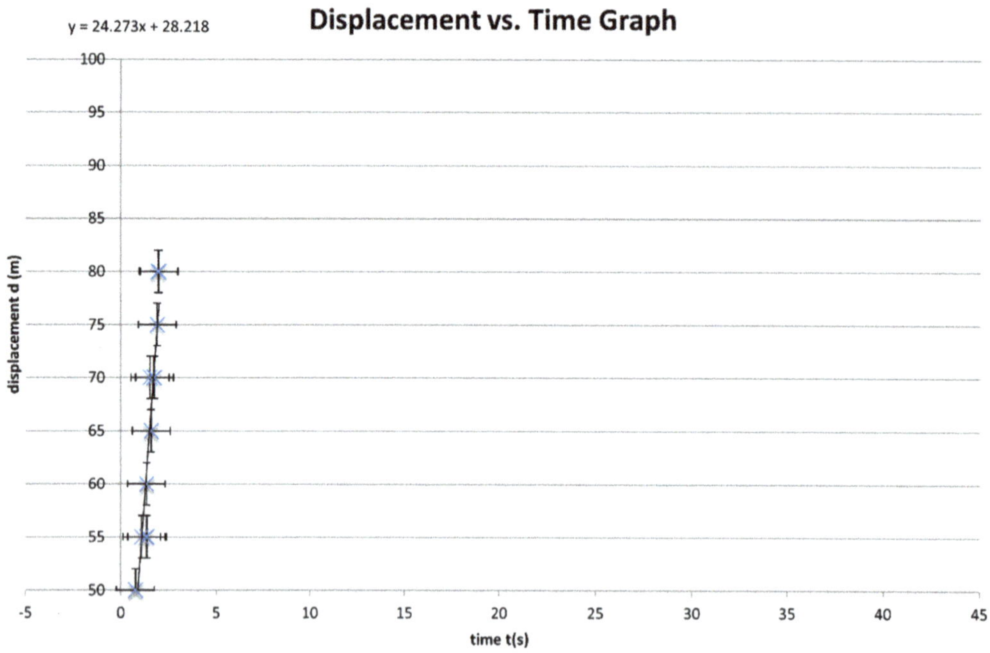

Figure 3-8 Poor space management in graphing.

To adjust the minimum and maximum values of the axis, double click on the values. A proper window will open.

The second possible problem is a bit more complex. As you remember from the first lab in this course, sometimes big mistakes are caused by human error. For example, misplacing a decimal point may cause a graph to appear as shown in Figure 3-9.

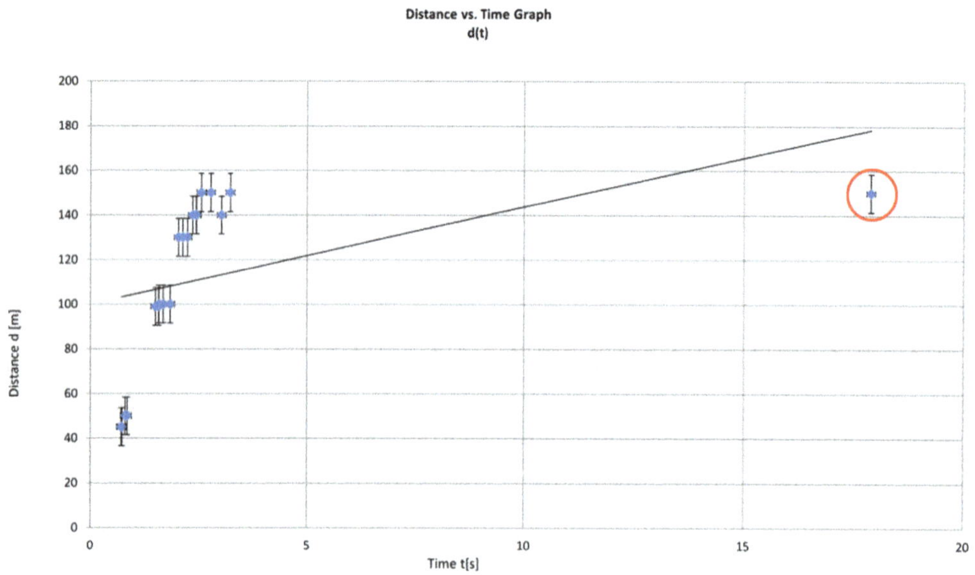

Figure 3-9 Dealing with outliers in graphing.

In that case, you may delete the point that differs so much. As you can see in Figure 3-9, leaving that point in might distort the parameters.

After removing outliers, Excel will automatically stretch the graph. If not, see the first answer.

That being said, be very careful when deleting data from your analysis. There may be good reasons for that one not fitting point to be so far on the graph. After removing certain points from analysis, make sure to explain that in the lab report.

Data Interpretation

Interpret the parameters obtained in the first graph by comparing them with the formula for the $d(t)$ function, when $\vec{v} = const$.

Equation of the best fit line as shown in the graph: ______________________________

Based on the equation, what is the velocity of the glider?

v = ______________________________

How do you interpret the free term? What is it?

Interpret the parameters obtained in the second graph by comparing them with the formula for the $d(t)$ function, when $\vec{a} = const$.

Equation of the best fit line as shown in the graph: ______________________________

Based on the equation, what is the acceleration of the glider?

a = ____________________

Based on the measurement of the angle of elevation in **Step Four**, calculate the angle of elevation of the air track.

$$a = g \cdot \sin\theta$$

a = ____________________

Compare the acceleration values (find the percent difference).

$$percent\ difference = \frac{|difference\ of\ two\ values|}{average\ of\ the\ same\ values} \cdot 100\%$$

Error Analysis

Compare both graphs. Is there any difference in variance of the points plotted in those graphs? Describe the difference. List possible reasons for the different variances.

What are the main sources of error in this experiment? What did you do to improve the experimental settings to avoid measurement errors?

Self-Reflection

Physicists quantify everything. Figure 3-10 shows a graph displaying students' involvement (the vertical axis) during lab (time is on the horizontal axis).

Which graph in Figure 3-10 represents your involvement during today's lab? If none of these reflects your view, please draw the graph yourself.

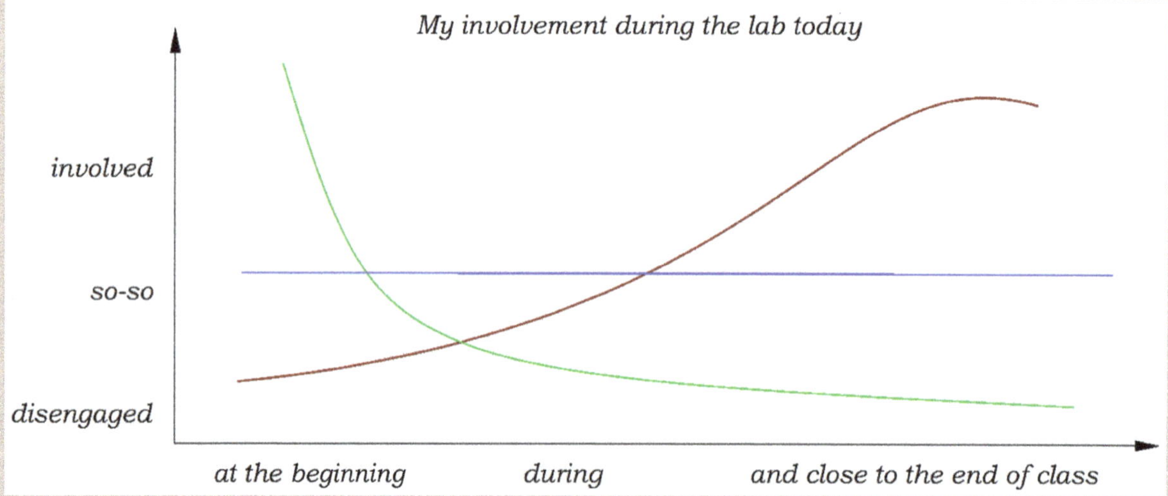

Figure 3-10 Graphical representation of students' involvement during lab.

Lab 4

Kinematic Graphs

Introduction

The purpose of this lab is to understand kinematic graphs and be able to describe motion based on information read in graphs.

Upon completion of this lab, students will

- Describe motion with graphs of distance, velocity, and acceleration as a function of time.
- Read the distance, displacement, velocity, and acceleration from various graphs.
- Reconstruct physical motion based on information provided by kinematic graphs.

Theory–Math Review

1. What is the equation of a linear function? Sketch a graph of a linear function.

f(x) =

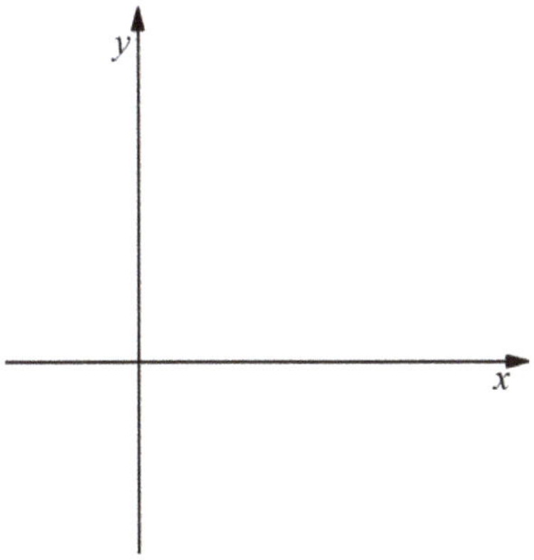

2. What is the equation of a quadratic function? Sketch a graph of a quadratic function.

f(x) =

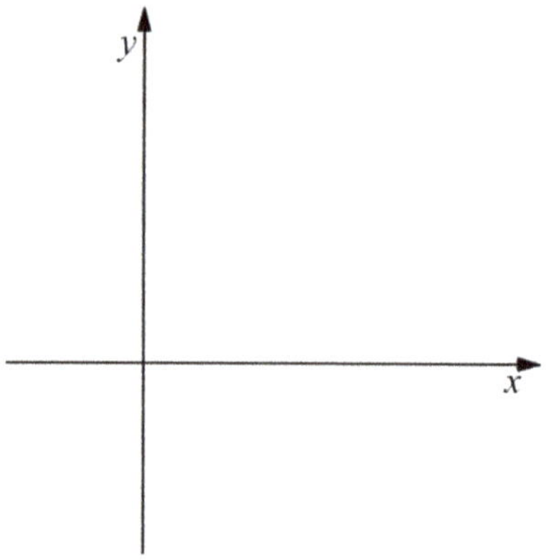

3. How do we recognize whether a function is increasing, decreasing, or constant?

Kinematic Graphs

There are three basic functions that are used to describe uniform and accelerated motion in kinematics: constant linear (Figure 4-1a), increasing linear (Figure 4-1b), and quadratic function (Figure 4-1c).

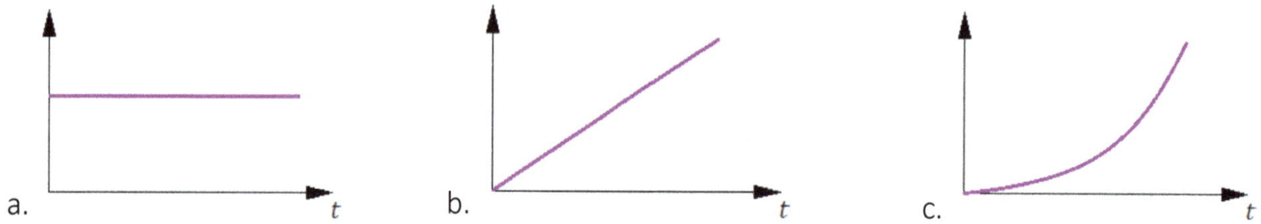

Figure 4-1 Basic kinematic graphs.

1. In Figure 4-1, vertical axes were not labeled by purpose. Focus only on the shape of the graph shown in the figure. Which graph represents
   - distance as a function of time for an object at rest? a b c
   - distance as a function of time in uniform motion? a b c
   - distance as a function of time in accelerated motion? a b c
   - velocity as a function of time in uniform motion? a b c
   - velocity as a function of time in accelerated motion? a b c
   - acceleration as a function of time in accelerated motion? a b c

Distance Versus Displacement

1. Define distance and displacement:

   Distance is ...

   Displacement is ...

2. Route 80 connects New York, NY, and San Francisco, CA. The length of Route 80 is approximately 2,900 mi. If you decide to fly from San Francisco to New York, the distance in the flight tracing monitor shows approximately 2,570 mi. Which value is closer to the definition of distance in physics? Which is closer to displacement as defined in physics?

Graphs in Uniform Motion

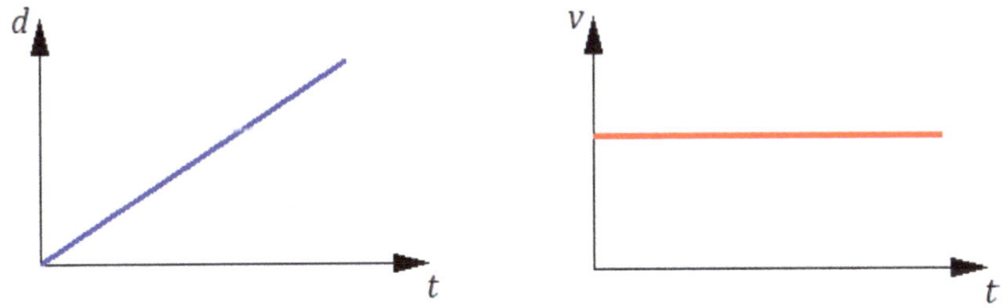

Figure 4-2 Distance and velocity as a function of time in uniform motion.

1. Define uniform motion based on the graphs in Figure 4-2. What do these graphs show about distance and velocity in uniform motion?

2. Distance can be read from the d(t) graph directly, as can velocity from the v(t) graph. Show how to find velocity on the d(t) graph and how to find distance on the v(t) graph.

3. While waiting for his advisor, a student is walking back and forth in the hall. After pacing back and forth twice, he stopped for a minute to have a look at the recently posted announcements. Show his movement on the following d(t) graph.

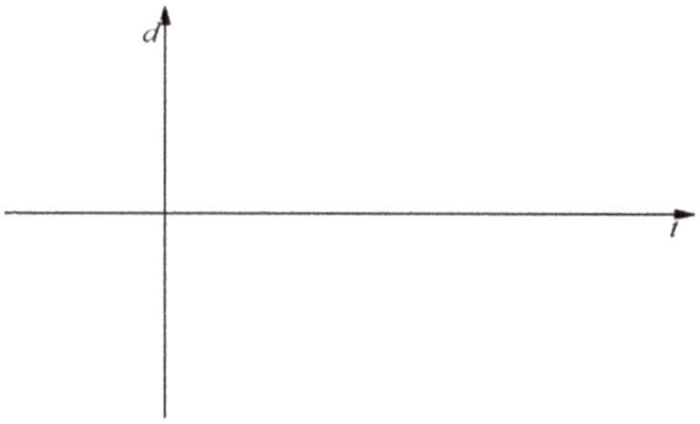

4. Describe the motion presented in Figure 4-3.

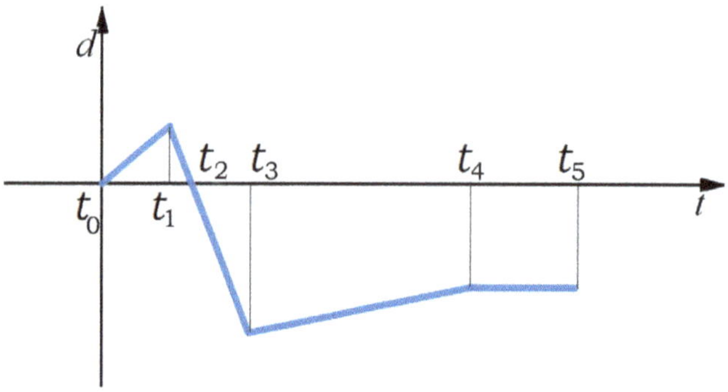

Figure 4-3 Distance as a function of time, $t_0 - t_5$.

$t_0 - t_1$: ______________________________

$t_1 - t_2$: ______________________________

$t_2 - t_3$: ______________________________

$t_3 - t_4$: ______________________________

$t_4 - t_5$: ______________________________

Graphs in Accelerated Motion

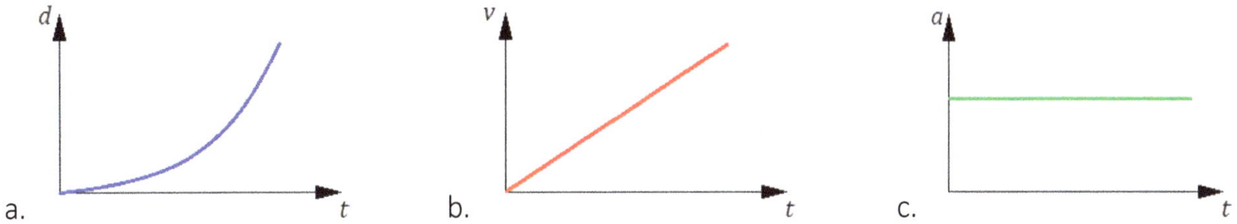

Figure 4-4 Distance, velocity, and acceleration as a function of time in accelerated motion.

1. Define accelerated motion based on the graphs in Figure 4-4. What do these graphs show about distance, velocity, and acceleration in uniform motion?

2. Distance, velocity, and acceleration can be read from the ***d***(*t*), ***v***(*t*), and ***a***(*t*) graphs directly. Explain or show in Figure 4-4 how to find
   - velocity on the ***d***(*t*) graph
   - acceleration on the ***v***(*t*) graph
   - distance on the ***v***(*t*) graph
   - velocity on the ***a***(*t*) graph

3. Figure 4-5 presents velocity as a function of time.

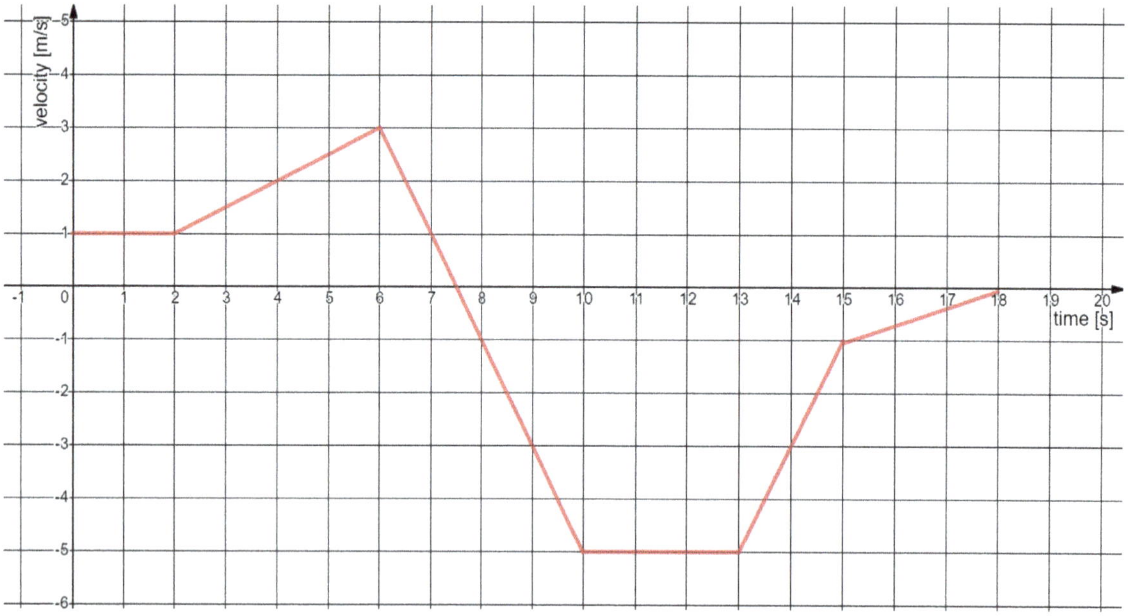

Figure 4-5 Velocity as a function of time.

Read from the graph acceleration at:

- $t = 1$ s $a =$ ______________
- $t = 3$ s $a =$ ______________
- $t = 7$ s $a =$ ______________
- $t = 9$ s $a =$ ______________
- $t = 11$ s $a =$ ______________
- $t = 14$ s $a =$ ______________
- $t = 16$ s $a =$ ______________

Based on the graph, calculate the distance traveled:

- between t_0 and t_2 $d =$ ______________
- between t_2 and t_6 $d =$ ______________
- between t_6 and $t_{7.5}$ $d =$ ______________
- between $t_{7.5}$ and t_{10} $d =$ ______________
- between t_{10} and t_{13} $d =$ ______________
- between t_{13} and t_{15} $d =$ ______________
- between t_{15} and t_{18} $d =$ ______________

4. Calculate total distance of motion presented in Figure 4-5.

distance = ______________________________

5. Calculate total displacement presented in Figure 4-5.

displacement = ______________________________

6. Figure 4-6 shows a $v(t)$ graph. Complete the $d(t)$ and $a(t)$ graphs for the same motion.

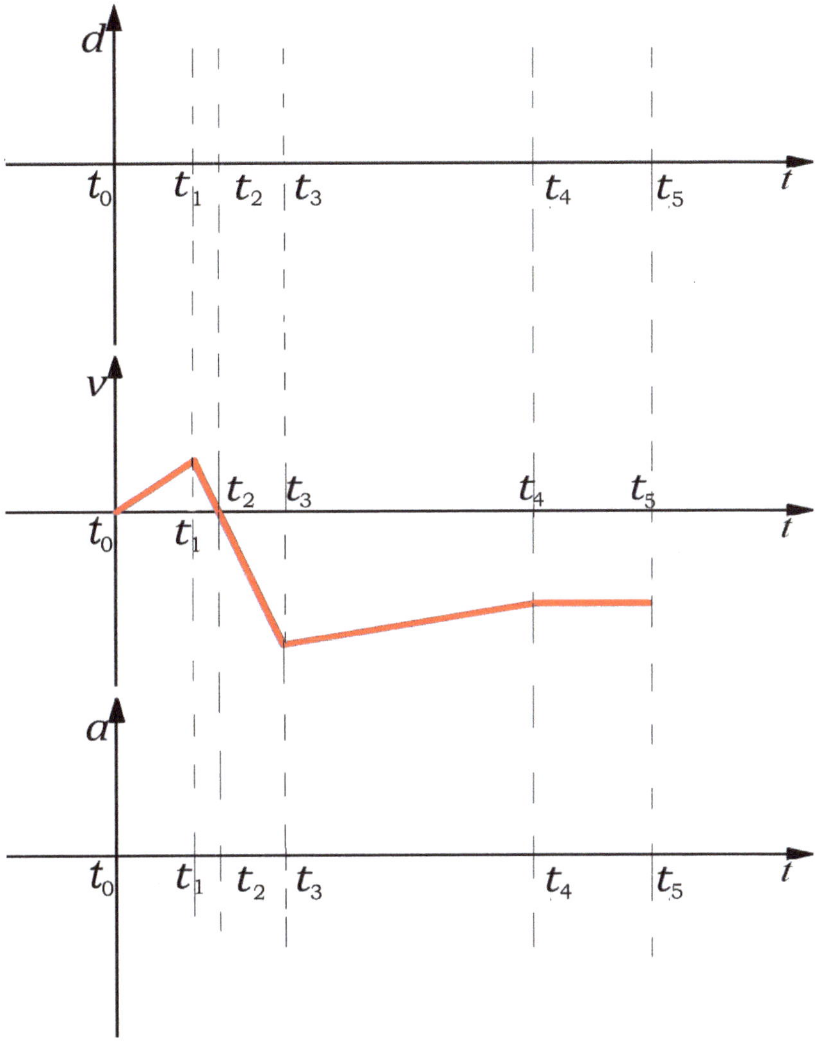

Figure 4-6 Velocity as a function of time graph.

Lab 5

Pendulum

Introduction

The purpose of this lab is to investigate the factors that might affect the period of a pendulum, namely the angle of the swing, the mass of the bob, and the length of the thread.

Upon completion of this lab, students will

- Know terms associated with harmonic motion and oscillations.
- Understand the procedure of isolating the effects of independent variables (elements of a controlled experiment).
- Explain how weight, angle of swing, and length affect the period of a pendulum.
- Analyze the terms in mathematical formula as they predict the results based on the initial conditions.
- Pose precise research questions and state hypotheses.
- Validate experimental results.
- Recognize confirmation bias and its sources.

Theory

A pendulum is a weight suspended on a thread. Swinging back and forth, it oscillates about the equilibrium position (Figure 5-1). In ideal settings, the thread should be inextensible (inelastic) and weightless. Since a thread that justifies such conditions does not exist, we use a thread whose weight is much smaller than the weight of a bob.

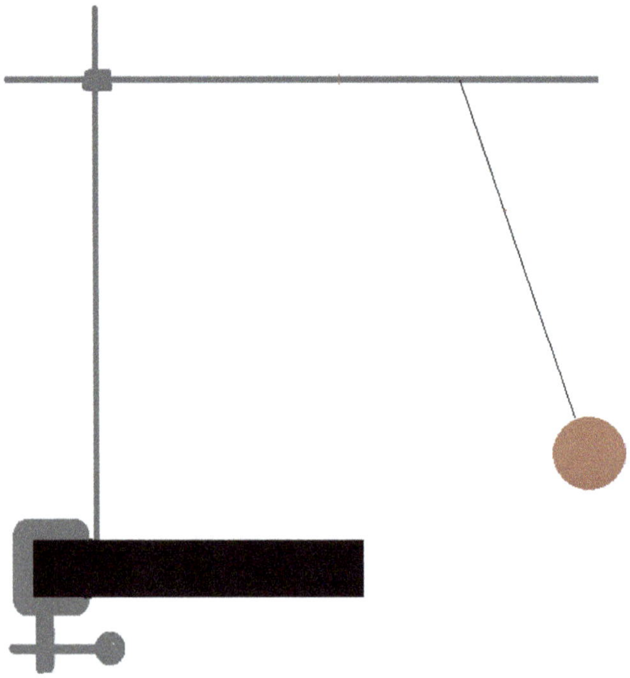

Figure 5-1 Pendulum.

Based on your textbook, define in your own words:

Oscillation:

Amplitude:

Period:

Frequency:

Part 1. Period of a Pendulum as a Function of the Starting Angle $T(\theta)$

In lab experiments, all three variables (the initial angle of swing, the mass, and the length) cannot be studied in one measurement. Thus, this experiment is planned to examine the variables one at a time, by performing three separate data collections (Part 1, 2, and 3 of this lab).

Critical Thinking

If we randomly change all three variables, the angle of swing, the mass of the bob, and the length of the pendulum, we will observe random effects on the period of the pendulum. How would we determine which variable influenced the period of the pendulum?

Experimental Method: Posing Questions

While designing lab experiments, it is critically important to know how to ask a question that can be answered by designing an experiment. In this lab, I propose the following question:

How does changing the angle of a pendulum affect its period?

The question indicates both investigated variables, the angle and the period. The angle is set by the researcher, and the assumption is that the period depends on it.

Complete the following sentence to state your prediction.

The ____________________ (larger/smaller) the initial angle,

the ____________________ (longer/shorter) the period.

Experimental Procedure

In this part of the lab, you will examine the relationship between the initial angle of the swing and the period of the pendulum. Thus, the mass of the bob and the length of the pendulum remain the same during the experiment.

Step-by-step procedure:

Step One. Starting at an angle indicated in Table 5-1, allow the pendulum to swing.

Step Two. Measure the pendulum's period for 10 oscillations using a stopwatch. You can repeat the measurement three times or have three group members taking the measurements independently. Record the times measured for 10 oscillations under t_1, t_2, and t_3 in Table 5-1.

Step Three. Find the average of t_1, t_2, and t_3 (the fifth column of the table).

Step Four. Find the period of the pendulum by dividing the average time by 10 (the last column of the table).

Step Five. Change the starting angle to 10°. Repeat.

Step Six. Find the average time for 10 oscillations and, finally, the period of the pendulum.

Table 5-1 Period of a Pendulum as a Function of the Starting Angle Data Collection Table.

Starting Angle	t_1 [s]	t_2 [s]	t_3 [s]	$t_{average}$	$T = t_{ave}/10$
<5°					
5°					
10°					
15°					

Critical Evaluation

Before stating your conclusion, estimate the measurement error. Since the human reaction time is usually estimated at 0.25 s, and in this experimental method the human reaction is a significant factor, the time measurement cannot be more precise than that. Also consider the variance of three independent measurements. Discuss with your group the precision of numeric values presented in the last column of the table. To how many decimal places should you round the results?

Observations/Conclusions

Does the conclusion confirm your prediction? Explain.

Part 2. Period of a Pendulum as a Function of the Mass of the Bob $T(m)$

In this part of the lab, you will examine the relationship between the mass of the bob and the period of the pendulum. Thus, the initial angle of the swing and the length of the pendulum remain the same.

Question

How does the mass of the bob affect the period of the pendulum?

Prediction/Hypothesis

Complete the following sentence to state your prediction.

The ____________________ (heavier/lighter) the bob,

the ____________________ (longer/shorter) the period.

Experimental Procedure

Repeat the experimental procedure described step-by-step in Part 1. This time, keep the starting angle the same, not larger than 5°. When adding extra weights to the bob, make sure that you do not change the length of the pendulum (the extra masses should not be hung below the bob; instead, add them on top of it). Remember the precision of the results presented in the last column of Table 5-2.

Table 5-2 Period of a Pendulum as a Function of the Mass of the Bob Data collection Table.

Mass [g]	t_1 [s]	t_2 [s]	t_3 [s]	$t_{average}$	$T = t_{ave}/10$
25 g					
50 g					
75 g					
100 g					

Observations/Conclusions

Does the conclusion confirm your prediction? Explain.

Part 3. Period of a Pendulum as a Function of the Length of the Pendulum $T(L)$

In this part of the lab, you will examine the relationship between the length of the pendulum and the period of the pendulum. Thus, the initial angle of the swing and the mass of the bob remain the same.

Question

How does the length of the pendulum affect its period?

Prediction/Hypothesis

Complete the following sentence to state your prediction.

The ____________________ (longer/shorter) the pendulum,

the ____________________ (longer/shorter) its period.

Experimental Procedure

Repeat the experimental procedure described step-by-step in Part 1. This time, keep the mass of the bob and the starting angle the same.

Repeat the measurement for at least four various lengths. When changing the length, make sure that the second length is about double the first one, the third length is about three times the first one, and the fourth length is about four times the original length. Collect the measurements in Table 5-3. Remember the precision of the results presented in the last column of the table.

Table 5-3 Period of a Pendulum as a Function of the Length of the Pendulum Data Collection Table.

Length [m]	t_1 [s]	t_2 [s]	t_3 [s]	$t_{average}$	T = $t_{ave}/10$
L_1 =					
L_2 =					
L_3 =					
L_4 =					

Observations/Conclusions

Does the conclusion confirm your prediction? Explain.

Part 4. Comparing the Experimental and Calculated Values

Calculate T using the lengths of the pendulum chosen by your team in Part 4 of this lab (Table 5-4) and the formula

$$T = 2\pi\sqrt{\frac{L}{g}}$$

Write down the calculated values in the second column in Table 5-4.

Copy the experimental values of the period of the pendulum from the last column of Table 5-3 to the third column in Table 5-4.

Use the formula

$$\frac{\left|T_{calculated} - T_{experimental}\right|}{\left(T_{calculated} + T_{experimental}\right)/2} \cdot 100\%$$

to find the percent difference between the experimental and calculated values (Column 4).

Table 5-4 Comparing the Experimental and Calculated Values.

Length [m] *Copy the first column from Table 5-3.*	$T_{calculated}$ *Calculate using the previously noted formula.*	$T_{experimental}$ *Copy the last column from Table 5-3.*	**Percent Difference** *Calculate using the previously noted formula.*
L_1 =			
L_2 =			
L_3 =			
L_4 =			

If the difference is greater than 5%, consider repeating the experimental procedure in Part 3.

If the difference is less than or equal to 5%, list the sources of error.

Discussion

Recall your predictions and observations.

Based on the formula, the period of the pendulum depends on the length of the pendulum L and gravitational acceleration g. The formula does not include mass; thus, the period of the pendulum does not depend on its mass. There are no trigonometric functions either, so changing angles does not change the period of the pendulum.[1]

- Did your predictions in Part 1 and Part 2 of this lab exercise reflect the theory presented in the formula for T?
- Did your conclusions reflect the theory presented in the equation in Part 4?

[1] For small angles, the change in the period of the pendulum depending on the initial angle is negligible. In this introductory physics lab, a small angle approximation is sufficient, especially considering the method of data collection.

Reflection

If your conclusions were not reflecting the theory, but simply confirming your predictions, you just experienced ***confirmation bias***. That happens to researchers in science laboratories, especially if a researcher comes to the lab with some preconceived notions about the expected results. That may impact data collection so strongly that the results are not objective anymore.

Here is a true story:

A student brings data collection for a review. The instructor asks for interpretation of the data collected in Column E, which consistently shows the value of 1.1, as shown in the following picture.

	A	B	C	D	E
1	45.4	130.142	48.4	44	1.1
2	11.0	26.889	22.0	20	1.1
3	29.4	82.347	38.5	35	1.1
4	49.4	142.242	50.6	46	1.1
5	36.1	102.245	42.9	39	1.1
6	72.3	210.809	61.6	56	1.1
7	27.9	77.709	37.4	34	1.1
8	13.9	35.561	25.3	23	1.1

Instructor: *So, do you see any pattern in the last column of this table?*

Student: *Yes, the values are increasing.*

Instructor: *Increasing? 1.1... 1.1... 1.1... They look the same to me....*

Student: *Yeah, they **look the same**, but there is **an invisible increase**.*

There is a saying that people tend to hear what they want to hear and see what they want to see. The previous story is a good example of holding such a strong conviction that the values should increase that the student saw an increase even though the numerical values were constant.

Confirmation bias is dangerous, especially in scientific investigations associated with large commercial grants. In some cases, scientists are expected to confirm certain results, and, with good intention, they put all their effort into doing so. Think what that means for large projects of developing and marketing new pharmaceuticals. Discuss in your group.

Although science should be objective and unbiased, it is performed by humans who are not perfect. One of the learning outcomes of this course is for you to become a responsible scientist who produces proper lab data collections and can state reliable conclusions. Be aware of the confirmation bias and be open to findings even if the data contradict your prediction.

Lab 6

Projectile Motion

Introduction

The purpose of this lab is to investigate and graph the range of a projectile as a function of the angle.

Upon completion of this lab, students will

- Derive formulas based on general theory.
- Decompose the velocity vector into horizontal and vertical velocity vectors.
- Analyze projectile motion in terms of the horizontal uniform and vertical accelerated components.
- Apply modeling to complex phenomena.
- Synthesize models to describe complex phenomena.

Safety Note

Use the safety goggles when the team is launching the projectile. Never direct the launcher toward anybody. Be mindful of the fragile objects such as lab glass or cell phones around.

Theory

1. Here are equations for the **horizontally launched projectile** motion.

	Distance	Velocity
Horizontal uniform motion	$x = v_x \cdot t$	$v_x = const.$
Vertical accelerated motion (free fall)	$y = -\frac{1}{2}gt^2 + y_0$	$v_y = -gt$

2. Use the previously noted equations to derive the formula for the time of flight in a horizontal projectile motion. Write down the formula here.

3. Here are equations for the motion of a **projectile launched at an angle**.

	Distance	Velocity
Horizontal uniform motion	$x = v_x \cdot t$	$v_x = const.$
Vertical accelerated motion (free fall)	$y = -\frac{1}{2}gt^2 + v_{y_0}t + y_0$	$v_y = -gt + v_{y_0}t$

4. Complete Figure 6-1 by drawing components of the velocity vector at points *A* and *B*. Compare the drawing with other students' work.

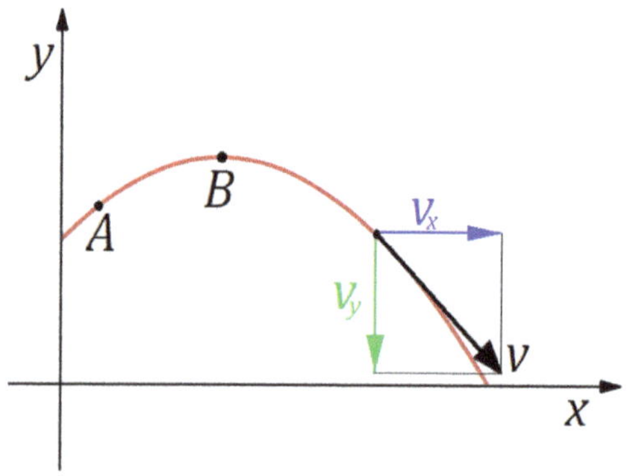

Figure 6-1 Components of the velocity vector in projectile motion launched at an angle.

5. Write formulas for components of the velocity vector v_x and v_y.

v_x = ______________________________

v_y = ______________________________

6. The previous equations describe the theory of a projectile launched at an angle. For practical purposes, the maximum height and the range as functions of the total initial velocity are more useful. Based on the theory presented in point 3 above, derive the formulas for the range of the projectile (show your work).

Part 1. Finding the Speed of the Projectile

In this part, you will use the properties of the horizontally launched projectile to find the speed of the projectile.

Step One. Set the projectile launcher as shown in Figure 6-2. Use the protractor to make sure the barrel is horizontal. Precision is a key here.

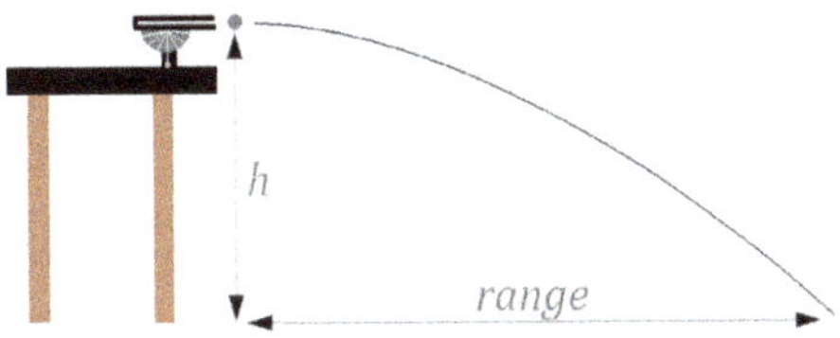

Figure 6-2 Horizontal projectile launcher settings.

Step Two. Measure the height h as shown in Figure 6-2.

h = __________________________________

Step Three. Use the measured height to calculate the time of flight.

$$t = \sqrt{\frac{2h}{g}}$$

t = __________________________________

Step Four. Shoot the projectile and measure the range of the flight. Write it down in Table 6-1, Column 2.

Step Five. Use the following formula to calculate the velocity. Write it down in Table 6-1, Column 3.

$$v = \frac{range}{time}$$

Table 6-1 Horizontally launched projectile data collection table.

Measurement	Range (measured)	Velocity (calculated)
Column 1	Column 2	Column 3
1		
2		
3		
4		
5		
6		
7		
8		
9		
10		

Step Six. Evaluate the data collection. Are there any outliers?

Step Seven. Find the average velocity.

$v_{average} =$ ______________________

Part 2. Range as a Function of the Launch Angle

In this part, you will investigate the range as a function of the launch angle of the projectile. Use the initial velocity generated in the launcher found in Part 1 of this exercise.

Before you begin data collection, hypothesize the results.

The optimal angle for the maximum range in projectile motion is ____________________.

Step One. Set the projectile launcher as shown in Figure 6-3. Use the protractor to find out the launch angle.

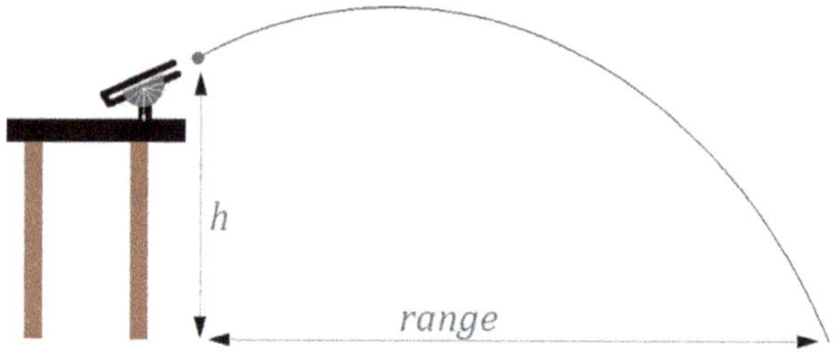

Figure 6-3 Projectile at an angle launcher setting.

Step Two. Set the launch angle at 10° above horizontal as indicated in Table 6-2, Column 2.

Step Three. Shoot the projectile. Use a measuring tape to measure the range of the flight. Record it in Table 6-2, Column 3.

Repeat Steps Two and Three for as many angles as you can in this experimental setting.

Step Four. Use the formula for the range of the projectile that you derived in the theory part, point 6, to calculate the range. Record the results in Column 4 of Table 6-2.

$$R = \frac{v_0^2}{g} \cdot \sin 2\theta$$

Step Five. Use the percent difference formula to compare the experimental and calculated results.

$$\frac{\left| R_{experimental} - R_{calculated} \right|}{R_{calculated}} \cdot 100\%$$

Table 6-2 Projectile launched at an angle data collection table.

Trial	Angle	Range (experimental)	Range (calculated)	Percent Difference
Column 1	Column 2	Column 3	Column 4	Column 5
1	10			
2	15			
3	20			
4	25			
5	30			
6	35			
7	40			
8	45			
9	50			
10	55			
11	60			
12	65			
13	70			
14	75			
15	80			

Data Evaluation

1. Is there any pattern in the data collected in Column 3?
2. Are there any outliers? Would you consider repeating measurements for those outliers?
3. Is the percent difference between the experimental and calculated results large (>5%)? Why?
4. What are the sources of experimental error? Could you eliminate them? How?

Data Analysis

Plot the data in the following graph. Put the angle on the horizontal axis and the range on the vertical one. Do the data points reflect the formula derived in the theory part, point 6 on page 55?

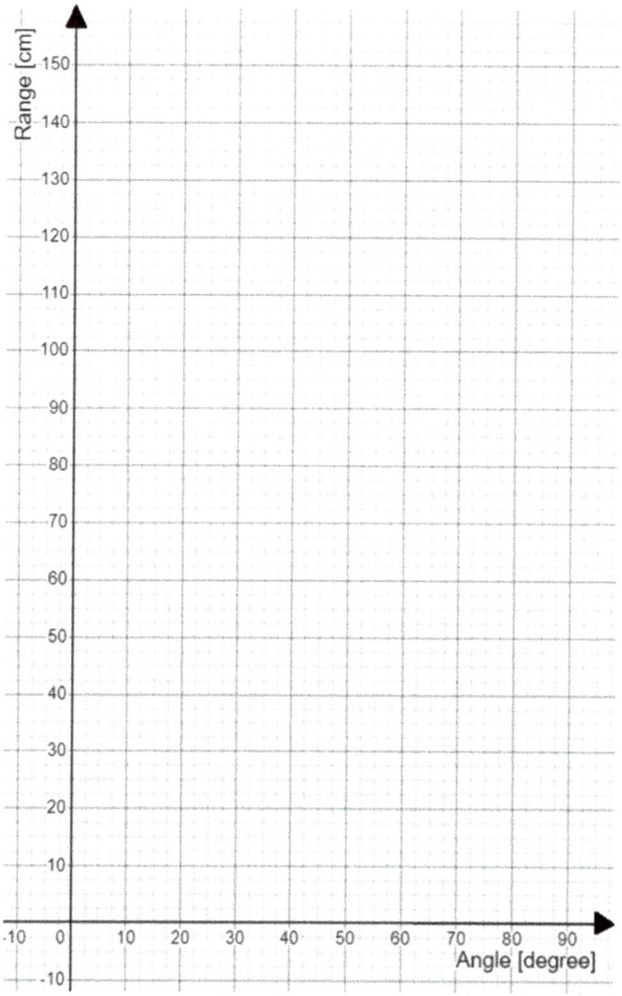

1. What is the maximum range based on the data collected in Table 6-2?

   Maximum range ______________________________

2. What angle corresponds with the maximum range?

   Corresponding angle __________________________

3. Was your prediction correct?

Reflection

Take some time to recall what happened in the lab today. Write down a few spontaneous thoughts.

How do you feel after class?

Lab 7
Force Table

Introduction

The purpose of this lab is to investigate the magnitude of the resultant vector as a function of mass and the angle between the component vectors.

Upon completion of this lab, students will

- Apply the graphical method to vector addition.
- Use the notion of vector to describe force.
- Add two vectors graphically.
- Apply the law of cosines to calculate the magnitude of the resultant.

Theory

Before you come to the lab, make sure that you can answer the following questions. Please refer to your textbook to find the answers.

1. What is force?

2. Give at least three examples of forces observed in nature.

3. Give three examples of vector quantities.

Math Window

Vectors are defined by their magnitude and direction. The magnitude is a numerical value expressed as the length of a vector. The magnitude of a vector whose coordinates are given can be found with the Pythagorean theorem. If the coordinates of vector $\vec{a}$ are [6,3], as shown in Figure 7-1, then the magnitude of vector $\vec{a}$ is

$$a = \sqrt{6^2 + 3^2} = \sqrt{45} = 6.71$$

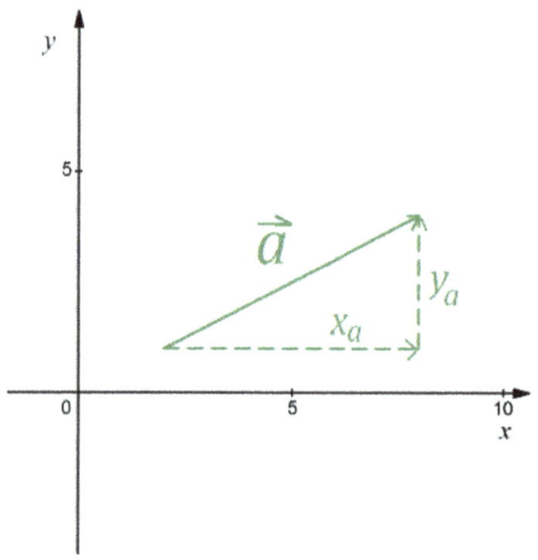

Figure 7-1 Vector $\vec{a}$ [6,3].

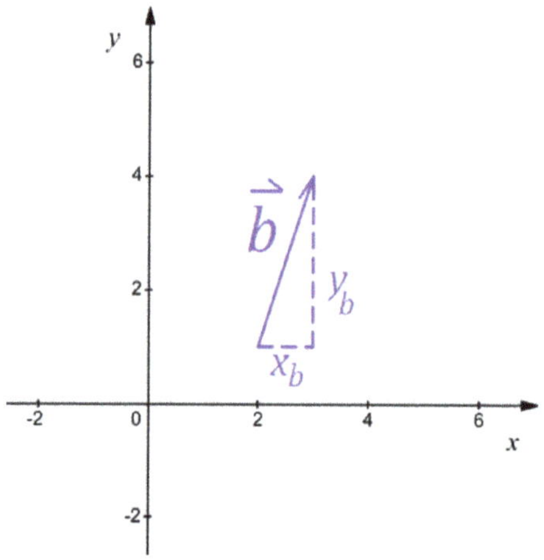

Figure 7-2 Vector $\vec{b}$ [1,3].

Find the magnitude of vector $\vec{b}$.

Vector Addition

Since vectors have two defining properties, magnitude and direction, both must be considered when adding vectors. Methods of vector addition include the coordinates method, the graphical method, and calculation based on the law of cosines.

Coordinates Method

If vectors are described by their coordinates, such as vectors $\vec{u}\,[x_u, y_u]$ and $\vec{v}[x_v, y_v]$, the resultant is defined as a vector whose coordinates are sums of the respective coordinates of vectors ***u*** and ***v***.

$$\overrightarrow{u+v}=[x_u+x_v, y_u+y_v]$$

To add vectors presented in Figure 7-2, $\vec{a}\,[6,3]$ and $\vec{b}\,[1,3]$, we add the respective coordinates:

$$\overrightarrow{a+b}=[6+1,3+3]=[7,6]$$

Find the magnitude of the resultant.

Graphical Method

In physics, it is convenient to use free body diagrams, in which vectors represent forces. Thus, the graphical method of vector addition might be useful. Consider two vectors, vector $\vec{a}$ and vector $\vec{b}$, as shown in Figure 7-3.

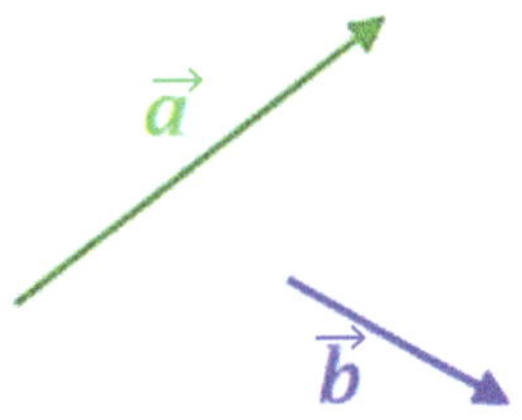

Figure 7-3 Vectors $\vec{a}$ and $\vec{b}$.

To graphically add the vectors, draw the two vectors head to tail (Figure 7-4). Then, connect the tail of the first vector and the head of the second vector to get the resultant $\vec{c}$ (Figure 7-5).

Figure 7-4 Vectors $\vec{a}$ and $\vec{b}$ head to tail.

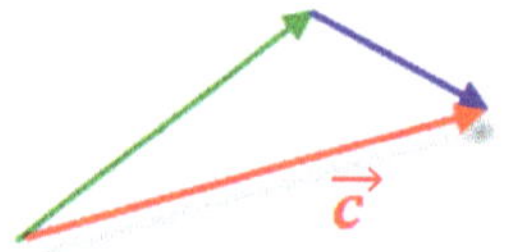

Figure 7-5 The resultant vector $\vec{c}$.

To find the magnitude of the resultant, draw both vectors in scale (for example, 1 cm represents one unit) and measure the resultant. Precision in drawing yields accurate results.

Notice that the magnitude of the resultant is smaller than the sum of the vectors' magnitudes and larger than their difference.

Calculation Method

One of the methods used to calculate the magnitude of the resultant is to apply the law of cosines,

$$c^2 = a^2 + b^2 - 2ab \cdot \cos\theta$$

where $\vec{a}$ and $\vec{b}$ are the vectors that are being added, θ is the angle between them as shown in Figure 7-6, and $\vec{c}$ is the resultant.

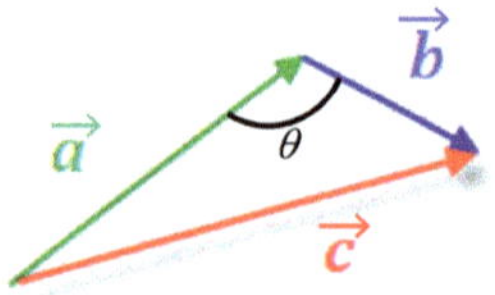

Figure 7-6 Vectors $\vec{a}$ and $\vec{b}$ and the resultant $\vec{c}$.

Exercise

1. Find the magnitude of the resultant vectors using the graphical method.
2. Calculate the magnitude of the resultant vector using the law of cosines.
3. Compare the results using the percent difference formula:

$$\frac{|\textit{difference between the measured and the calculated value}|}{\textit{average of the two values}} \cdot 100\,\%$$

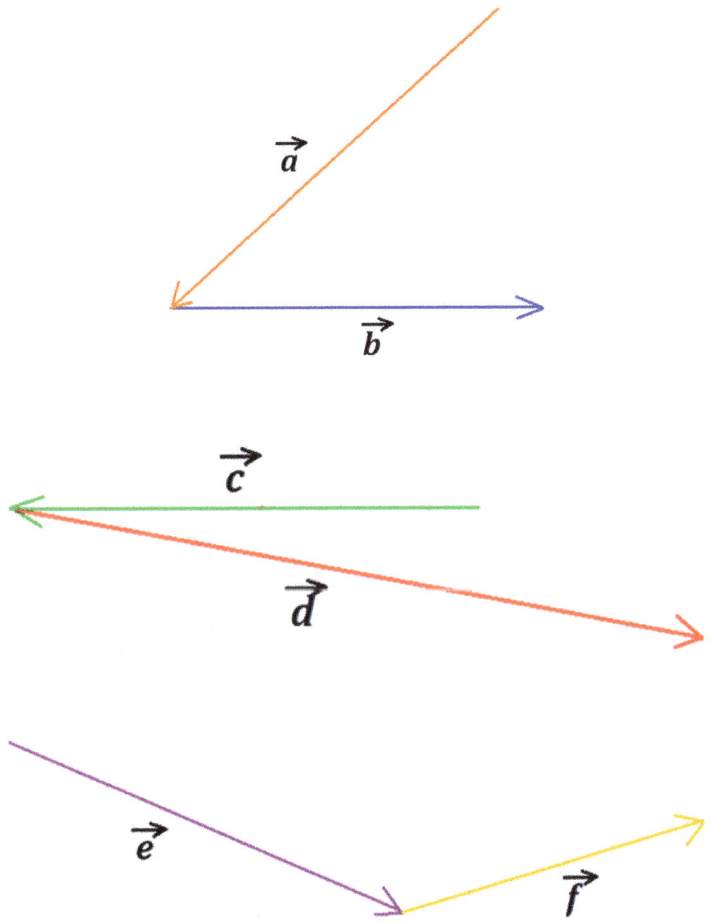

Vector 1	Vector 2	Resultant Measured	Resultant Calculated	Percent Difference
$\lvert\vec{a}\rvert =$	$\lvert\vec{b}\rvert =$	$\lvert\overrightarrow{a+b}\rvert =$	$\lvert\overrightarrow{a+b}\rvert =$	
$\lvert\vec{c}\rvert =$	$\lvert\vec{d}\rvert =$	$\lvert\overrightarrow{c+d}\rvert =$	$\lvert\overrightarrow{c+d}\rvert =$	
$\lvert\vec{e}\rvert =$	$\lvert\vec{f}\rvert =$	$\lvert\overrightarrow{e+f}\rvert =$	$\lvert\overrightarrow{e+f}\rvert =$	

One-Minute Mindfulness Exercise

Stop for a minute and give it some thought.

1. Which method of vector addition do you prefer?
2. Recall the math class in which you learned how to use a ruler, use a protractor, and measure the length of a segment. You learned how to speak math to do science.
3. Go online and find a good app for vector addition. In this lab, we would prefer an app that uses the magnitudes of two vectors and the angle between them rather than vector coordinates. If you cannot find such an app, maybe this is the time to build it yourself.

Experimental Settings

1. Fix the pulleys on the force table as shown in Figure 7-7.
2. Level the table.
3. Put a marble anywhere on the force table. If the marble does not roll, the table is leveled. If it does, then use the adjustable legs to level the table ("The marble ball test").
4. Fasten two strings to the ring and secure the ring over the center of the table using a pen or pencil.
5. Attach the same weights to the strings. Use pulleys to minimize friction (Figure 7-7).

Figure 7-7 Force table; (a) side view, (b) top view.

b.

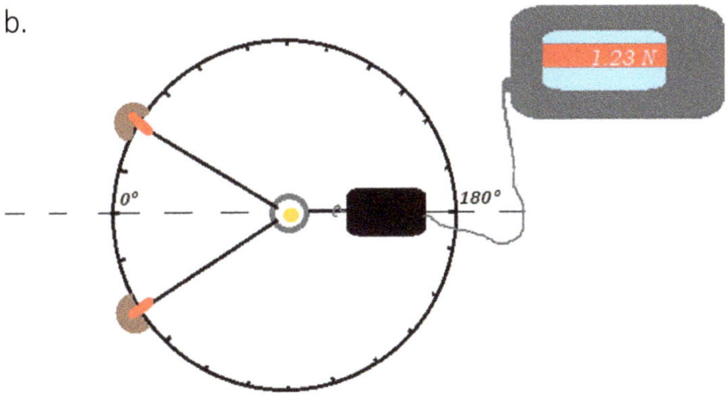

Figure 7-7 (*continued*)

6. Place the force sensor at 180° (Figure 7-7b).
7. Pull the ring with the force sensor so the pencil is exactly in the center of the ring. That means that the forces are in equilibrium.
8. Notice that the force sensor measures equilibrant, not resultant. Equilibrant and resultant are opposite in direction but the same in magnitude.

Part 1. Finding the Resultant on the Force Table–Fixed Angles

In this part of the lab, you will investigate the magnitude of the resultant as a function of mass

Step-by-step procedure:

Step One. Place one pulley at 40° and the second one at 320°.

Step Two. Notice that angle θ between 40° and 320° is 80°.

Step Three. Add the same mass to both pulleys as shown in Table 7-1.

Step Four. Calculate the weight on the pulleys. Remember to convert *g* to *kg* before calculating weight:

$$weight = mass \cdot 9.81 \frac{m}{s^2}$$

Step Five. Calculate the resultant using the law of cosines:

$$resultant = weight \cdot \sqrt{2\left(1+\cos 80^\circ\right)}$$

Note the results in the third column of Table 7-1.

Step Six. Zero the sensor before measuring.

Step Seven. Find the equilibrant using the force sensor. Note the results in the fourth column in Table 7-1.

Step Eight. Compare the results by finding the percentage difference.

$$percent\ difference = \frac{|difference\ of\ the\ two\ values|}{average\ of\ the\ two\ values} \cdot 100\%$$

Table 7-1 Comparison of the Resultant and Equilibrant (Fixed Angles).

Mass 1 & 2	Weight (calculated)	Resultant (calculated)	Equilibrant (measured with the force sensor)	Percent Difference
100 g				
150 g				
200 g				
250 g				
300 g				
400 g				

Critical Evaluation of the Results

What are your observations? Is there anything unexpected/surprising?

Is the percent difference between the measured and the calculated results always the same? Is there any pattern in the percent difference value; for example, increasing or decreasing with the increasing mass?

Part 2. Finding the Resultant on the Force Table–Fixed Mass

In this part of the lab, you will investigate the magnitude of the resultant as a function of the angle between the component vectors.

Step-by-step procedure:

Step One. Add the same mass to both pulleys.

Step Two. Calculate the weight suspended from each pulley. Remember to convert *g* to *kg* before calculating weight.

$$weight = mass \cdot 9.81\frac{m}{s^2}$$

Weight = ________________________________

Use this weight as a magnitude of a vector in each calculation.

Step Three. Place one pulley at 10° and the second one at 350°.

Step Four. Find angle θ between the pulleys (the angle between 10° and 350° is 20°).

Step Five. Calculate the resultant using the law of cosines:

$$resultant = weight \cdot \sqrt{2(1+\cos\theta)}$$

Note the results in the fourth column in Table 7-2.

Step Six. Find the equilibrant using the force sensor placed at 180°. Note the results in the fifth column in Table 7-2.

Step Seven. Compare the results by finding the percent difference.

Table 7-2 Comparison of the Resultant and Equilibrant (Fixed Masses).

Pulley 1	Pulley 2	Angle θ	Resultant (calculated)	Equilibrant (Force Sensor)	Percent Difference
10°	350°	20°			
20°	340°				
30°	330°				
40°	320°				
50°	310°				
60°	300°				

Critical Evaluation of the Results

What are your observations? Is there anything unexpected/surprising?

Is the percent difference between the measured and the calculated results always the same? Is there any pattern in the percent difference value; for example, increasing or decreasing with the increasing mass?

Where are the sources of experimental error in this exercise?

One-Minute Mindfulness Exercise

Congratulations, you are almost finished.

1. What did you learn today?

2. What was the funniest moment in class?

3. How was the group work today?

Lab 8

Inclined Plane

Introduction

The purpose of this lab is to examine the forces acting on an object on an inclined plane.

Upon completion of this lab, students will

- Draw a free body diagram.
- Recognize the forces acting on an object on an inclined plane.
- Use trigonometric functions to calculate the magnitudes of components of the force.
- Decompose the forces on an inclined plane.

Theory

1. A free body diagram (FBD) is a schematic picture used in physics that **does not include unnecessary details** (well-illustrated in Figure 8-1a) and allows examining forces acting on the body. The ability to draw FBDs is critical in modeling and problem-solving.

   Name the forces shown in the FBD in Figure 8-1b.

Figure 8-1 Comparison of a descriptive picture (a) and a free body diagram (b).

2. Draw an FBD of (a) a person sitting in a chair, and (b) a coffee mug standing on a table.

(a) (b)

Do your FBDs look similar?

3. The decomposition of a vector is opposite to vectors addition. We are looking for components of a vector: two vectors which, when added together, give the original one.

Any vector in the Cartesian plane can be decomposed into its *x* and *y* components (see Figure 8-2). Sometimes it is more convenient to use different directions for the decomposition of forces. An example of that is the decomposition of force on an inclined plane (Figure 8-3).

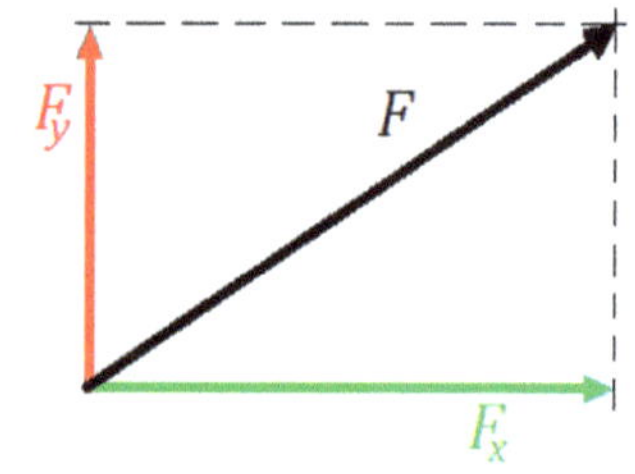

Figure 8-2 Components of force in the Cartesian plane.

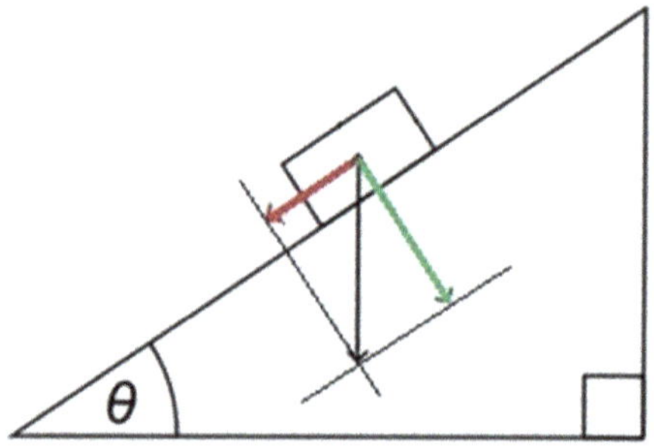

Figure 8-3 Components of force on an inclined plane.

4. How to draw an FBD on an inclined plane?

The gravitational force exerted on a block sitting on an inclined plane is represented by a vector in the FBD in Figure 8-4a.

The first step is to draw the parallel component (the red one in Figure 8-4b) and perpendicular vector (the green one in Figure 8-4c).

To calculate the magnitude of the components, first identify three similar right triangles and angle θ (Figure 8-4d).

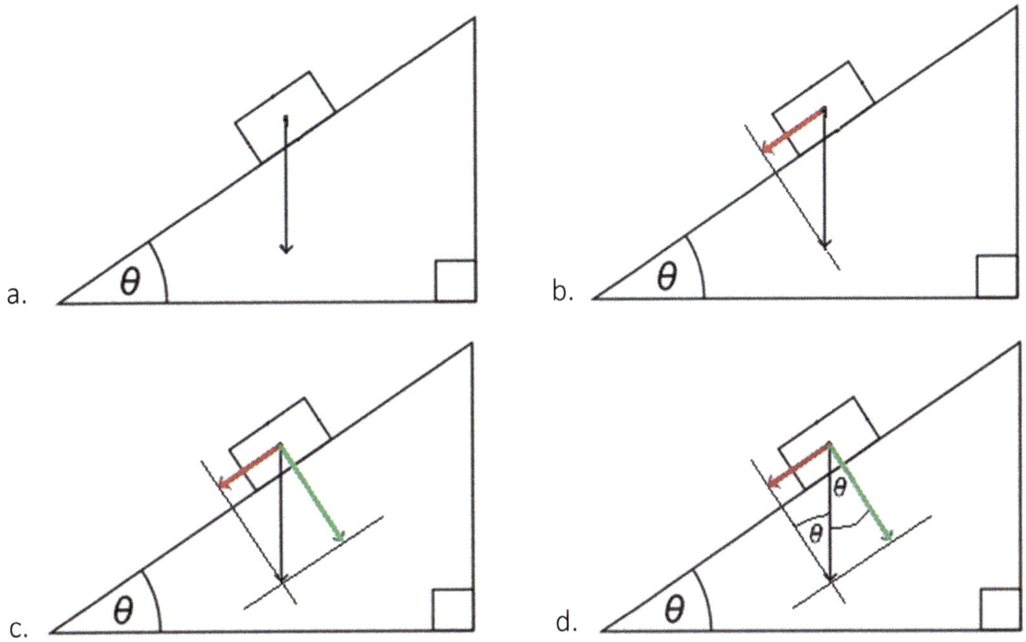

Figure 8-4 Decomposition of forces on an inclined plane, step-by-step tutorial.

5. How to calculate the magnitudes of vector components?

   By definition

$$\frac{F_{\parallel}}{F} = \sin\theta$$

so the parallel component can be expressed as

$$F_{\parallel} = F \cdot \sin\theta$$

Similarly, the perpendicular component can be expressed as

$$F_{\perp} = F \cdot \cos\theta$$

6. The components of weight are equilibrated by the reaction forces, frictional force and normal force (the contact force exerted on an object by the board), as shown in Figure 8-5

$$|F_{\parallel}| = |f|$$

and

$$|F_{\perp}| = |N|$$

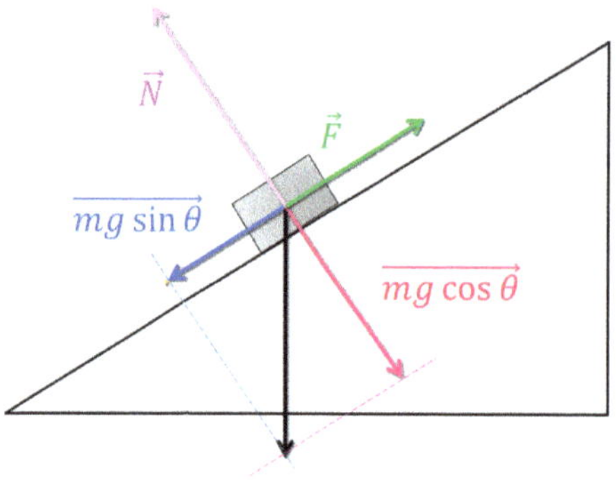

Figure 8-5 Reaction forces on the inclined plane.

Decomposition of Forces Exercise

Complete the following diagram. Show all forces acting on the object.

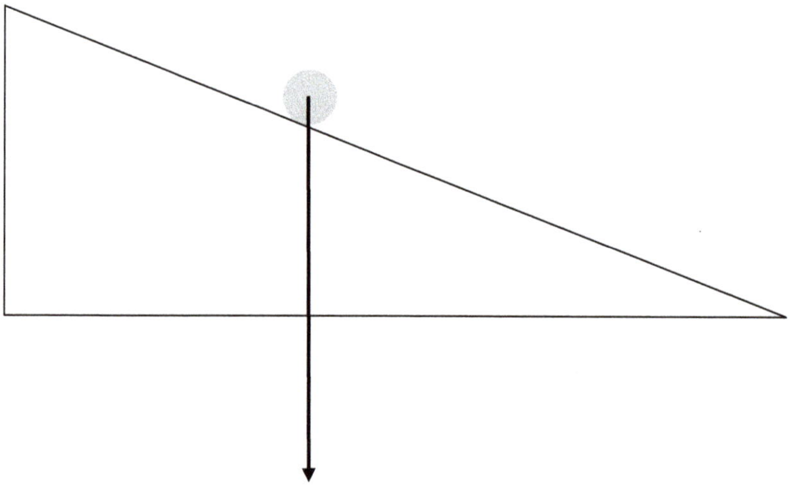

Grading Rubric for the Decomposition of Forces Exercise

Vector	Magnitude	Direction	Auxiliary Lines
Parallel component			
Friction			
Perpendicular component			
Normal force			
Total			

E–Expert (2 pts), R–Room for improvement (1 pt), N–Not acceptable (0 pts).

+ one extra point if nice and neat = ______________________________ out of 25 pts.

Experimental Procedure

In this part of the lab, you will investigate the parallel component of the weight vector.

Experimental procedure step-by-step:

Step One. Set the inclined plane as shown in Figure 8-6.

Step Two. Place the force sensor on the inclined plane. Attach a mass to the force sensor. Use a little cart to eliminate friction.

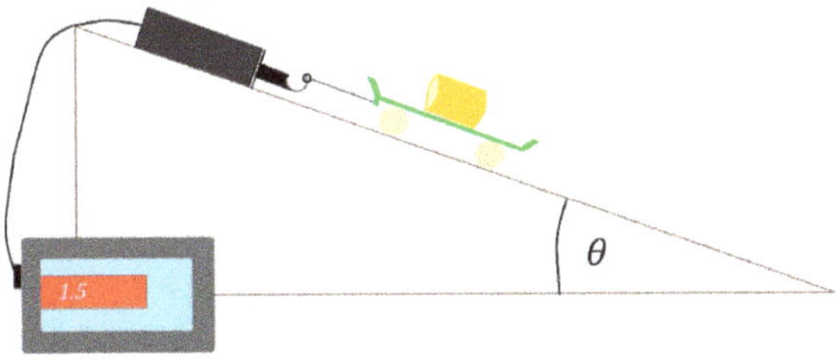

Figure 8-6 Inclined plane experimental settings.

Step Three. Design a method to measure the angle of inclination. Describe or sketch your method in the space that follows.

Step Four. Data collection.

1. Set the angle of inclination equal to 10°. Read the magnitude of the parallel component of the weight displayed on the force sensor.
2. Repeat the procedure for all angles in Table 8-1.

Table 8-1 Inclined plane data collection.

Trial	Angle of Inclination	Force Sensor Reading	Calculated Force	Percent Difference
1	10°			
2	15°			
3	20°			
4	25°			
5	30°			
6	35°			
7	40°			
8	45°			
9	50°			
10	55°			
11	60°			
12	65°			
13	70°			
14	75°			
15	80°			

Critical Evaluation

1. Are there any outliers in the experimental data column?
2. Evaluate the data collected in this exercise by calculating the percent difference (by now, you may know the formula by heart).
3. Is the percent difference <5% for all measurements? For which angles it is larger?
4. Is there any pattern in calculated percent difference? Is the accuracy increasing with the increasing angle?
5. Is the reading consistent? Is there any visible pattern in the collected data?

Notes:

Data Analysis

1. Create a graph of force as a function of the angle of inclination, $F = F(\theta)$. You can use Desmos or Excel.
2. What mathematical function can be used to model the plotted points? Why?

Present your findings to the instructor.

One-Minute

Mindfulness Exercise

Take some time to recall what happened in the lab today.

- Was there anything surprising in this lab?
- What was easy/difficult in this lab?
- What part of this lab will you remember after three years?
- How would you evaluate your group work?

Write down a few spontaneous words or phrases that come to mind.

Lab 9

Friction Coefficient

Introduction

The purpose of this lab is to investigate the frictional force as a function of mass.

Upon completion of this lab, students will

- Explain the phenomenon of friction.
- Differentiate static and kinetic friction.
- Improve lab techniques.
- Perform analysis of nonlinear graphs.

Theory

1. Explain the sources of friction.

2. Define friction coefficient.

3. Note how to calculate the coefficient of friction.

One-Minute Mindfulness Exercise

Before coming to class, quiet your cell phone. Be still for 10 sec.

Notice your colleagues and say something nice to greet them. Do not hesitate to begin the conversation, and show interest in others.

We need each other. We are in this class together.

Experimental Setting

In this part of the lab, you will determine the friction coefficient between wood and wood.

Step One. Connect the force sensor and the wooden block. Follow the diagram in Figure 9-1. Switch the force sensor to a graph view and start data collection by pressing the green arrow in the lower left corner. Gently pull the force sensor.

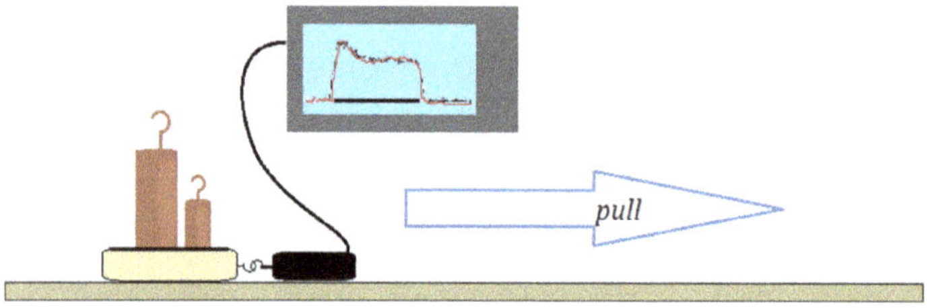

Figure 9-1 Friction coefficient experimental setting.

Step Two. Before attempting data collection, take time to learn how to record the force. Perform a few trials until you get a graph like the one shown in Figure 9-1. Experience tells us that it might take a while before you get similar readings.

Step Three. Learn how to read the graph. First, notice that there are two distinct areas in the graph, the peak and the plateau (Figure 9-2).

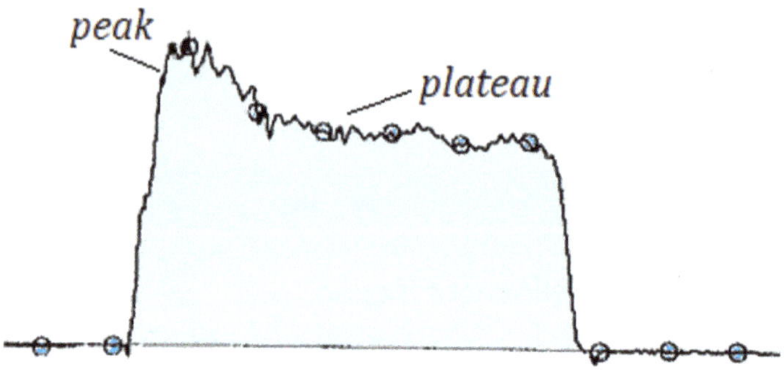

Figure 9-2 Pulling force needed to overcome static (peak) and kinetic (plateau) friction.

The peak is recorded when the motion was initiated. The maximum value indicates the static friction that was overcome to put the wooden block in motion.

The plateau is recorded later when the pulling force equilibrates the kinetic friction. Notice that kinetic friction is always less than the static one. To find the static friction, record the average force over the plateau area.

Part 1. Wood–Wood Friction

In this part of this lab, you will investigate the frictional force as the function of mass.

Step One. Place weights indicated in Column 1 of Table 9-1 on the block and record the measurements in Table 9-1.

Step Two. Record the total mass of the wooden block and weights ($m_w + m_b$ in Column 2). Calculate the normal force and record it in Column 3.

$$N = (m_w + m_b)kg \cdot 9.81\frac{m}{s^2}$$

Step Three. Read the frictional forces from the graph and record them in Column 4 and Column 5 of Table 9-1.

Table 9-1 Wood–wood friction data collection.

m_w [kg]	m_w+m_b [kg]	mg [N]	Static Friction	Kinetic Friction
Column 1	**Column 2**	**Column 3**	**Column 4**	**Column 5**
0.5				
0.6				
0.8				
1.0				
1.2				
1.4				
1.6				
1.8				
2.0				

Data Collection Evaluation

1. Are there any patterns in the collected data?
2. Are there any outliers?
3. Do your data make sense? How do you judge it?
4. Do you need to redo any of the measurements?

Discuss the answers to these questions with your team. Write down your observations.

Data Analysis

Plot data from Table 9-1 in two separate series. Plot the normal force ***N*** on the horizontal axis and the frictional force ***f*** on the vertical axis.

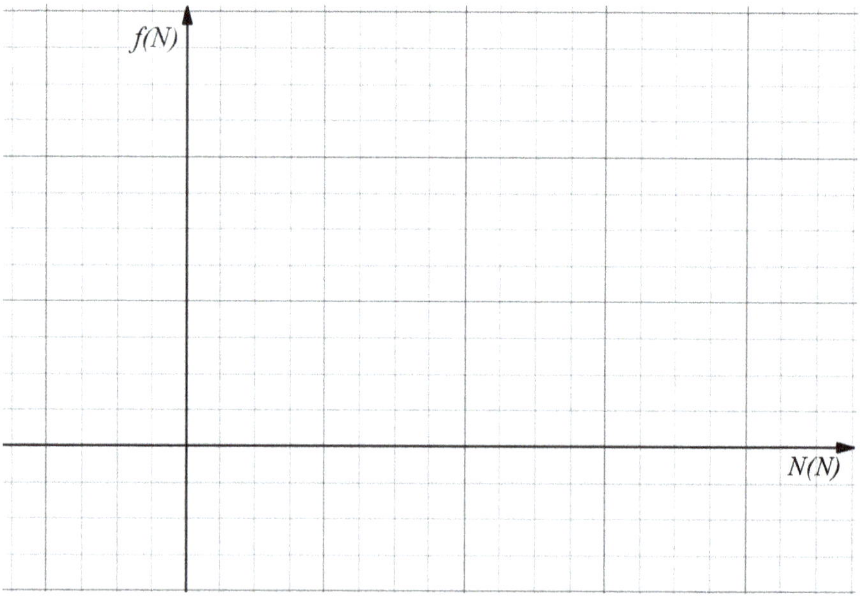

Read the friction coefficients from the graph.

f_s = ____________________

f_k = ____________________

Part 2. ______________________________ Friction

Choose other two surfaces, for example glass and cardboard or wood and metal, for this exercise. Fill out the title accordingly.

Repeat all steps from Part 1 for another pair of surfaces and record the values in Table 9-2.

Table 9-2 ________________ friction data collection.

m_w [kg]	m_w+m_b [kg]	mg [N]	Static Friction	Kinetic Friction
Column 1	Column 2	Column 3	Column 4	Column 5
0.5				
0.6				
0.8				
1.0				
1.2				
1.4				
1.6				
1.8				
2.0				

Data Collection Evaluation

1. Are there any patterns in the collected data?
2. Are there any outliers?
3. Do your data make sense? How do you judge it?
4. Do you need to redo any of the measurements?

Discuss the answers to these questions with your team. Write down your observations.

Data Analysis

Plot data from Table 9-2 in two separate series. Plot the normal force ***N*** on the horizontal axis and the frictional force ***f*** on the vertical axis.

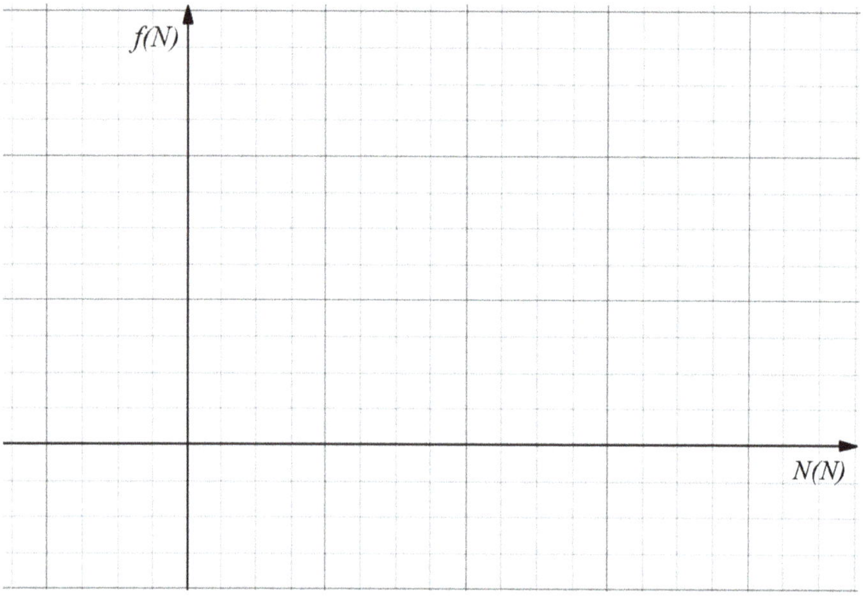

Read the friction coefficients from the graph.

f_s = ______________________________

f_k = ______________________________

Part 1 & 2 Conclusion

Compare your friction coefficient values determined during this lab session with a published table. Discuss the results.

Part 3. Explorations

1. How would you design an experiment to investigate the frictional force as a function of the surface area?

2. How would you design a lab experiment to investigate the friction coefficient as a function of humidity or precipitation?

Lab 10
Collisions

Introduction

The purpose of this lab is to investigate linear momentum during elastic and inelastic collisions.

Upon completion of this lab, students will

- Define elastic and inelastic collisions.
- Calculate the linear momentum of a system.
- Evaluate collision parameters based on the analysis of outcomes.
- Predict the outcome of a collision depending on masses and velocities of objects.

Theory

1. Define linear momentum.

2. State the law of conservation of linear momentum.

3. What is the difference between elastic and inelastic collision?

Real-Life Connection

After two cars collided, various parts were found around as shown in Figure 10-1. Based on that information, is there a way to assess which driver was driving faster? Discuss with your group.

Figure 10-1 Two cars collided.

Experimental Settings

1. Air track

   Set the air track as shown in Figure 10-2. Make sure that the air track is level before you start. If it is not, then level it.

Leveling the Air Track

Place a glider in the middle of the air track. Turn the air source on. If the glider is moving in any direction, slightly elevate the leg to which the glider is moving toward.

Take time to level the air track. The success of this experiment depends on this step.

2. Gliders

   In this experiment, you will use two identical gliders. Weigh the gliders.

   m_1 = ________________ m_2 = ________________

   Measure the length of the flags.

   L_1 = ________________ L_2 = ________________

   Notice that gliders have two bumpers: (1) a spring to model elastic collisions and (2) a Velcro to model inelastic collisions. Remember to set them properly for each part of this exercise.

3. Photogates

Arrange the photogates as shown in Figure 10-2. Set the photogates so that you can read the velocity of the glider.

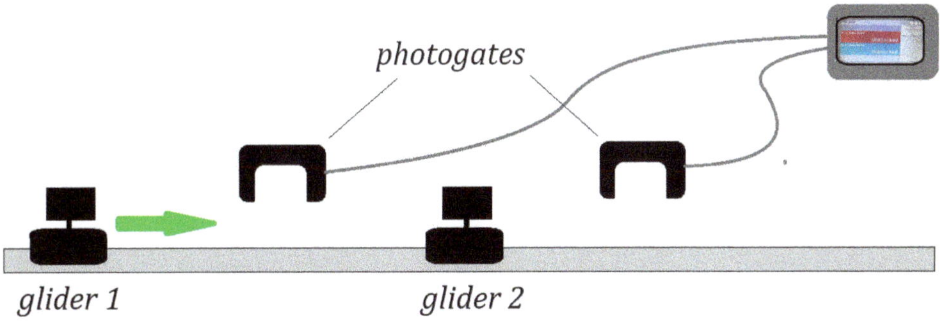

Figure 10-2 Photogates and gliders set on the air track.

4. Trial run

Set the gliders as shown in Figure 10-2. Push the first glider so it collides with the second one. Observe the second track. The expected scenario: Glider 1 passes under the photogate and collides with glider 2. After the collision, glider 1 stops and glider 2 passes under the other photogate.

5. Planning the experiment

Various scenarios for collisions include changing masses of either one of the gliders and changing the gliders' initial velocities. Such a complex set of initial conditions of the experiment requires planning.

The following diagrams show an overview of this lab exercise. The diagrams show the settings "before" the collision.

Part 1

Elastic collision

$m_1 = m_2$

$v_1 > 0, v_2 = 0$

Diagram 1

Part 2

Elastic collision

$m_1 > m_2$

$v_1 > 0, v_2 = 0$

Diagram 2

Part 3

Elastic collision

$m_1 < m_2$

$v_1 > 0, v_2 = 0$

Diagram 3

Part 4
Elastic collision
$m_1 = m_2$
$v_1 > 0, v_2 > 0$

Diagram 4

Part 5
Inelastic collision
$m_1 = m_2$
$v_1 > 0, v_2 = 0$

Diagram 5

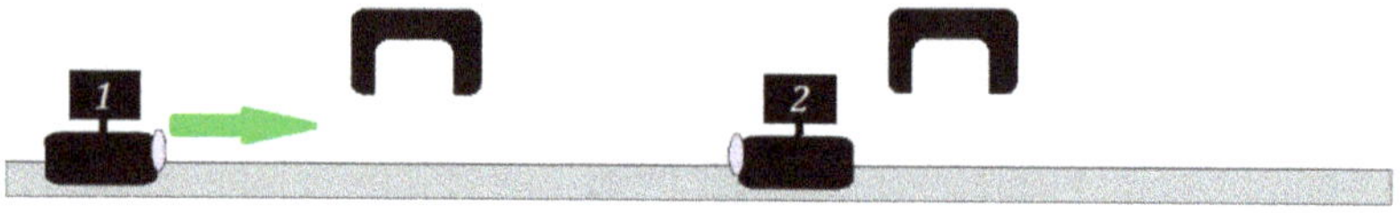

Part 1. Elastic Collisions, $m_1 = m_2$

In this part of the lab, you will test collisions between gliders of identical masses with a glider in motion colliding with the second one at rest (see Diagram 1).

Prediction Based on the Theory

According to the law of conservation of linear momentum,

$$p_{before} = p_{after}$$

Since

$$p = m \cdot v$$

the law of conservation of linear momentum adapted for this lab exercise looks like the following:

$$m_1 v_1 + m_2 v_2 = m_1 u_1 + m_2 u_2$$

where v stands for velocity before the collision, and u stands for the velocity after the collision.

In this lab setting, $m_1 = m_2$, so we can eliminate mass from the equations by dividing both sides by m:

$$v_1 + v_2 = u_1 + u_2$$

In addition, the second glider is at rest, $v_2 = 0$; thus, the equation becomes even simpler:

$$v_1 = u_1 + u_2$$

We decide the velocity of the first glider by putting it in motion. According to the previous equation, the velocity should be distributed somehow between u_1 and u_2. In the next few trials, you will observe the transfer of linear momentum when the gliders collide. Notice the distribution of momentum after the collision.

Testing

Place the first glider at the beginning of the air track and the second one in the center. The second glider should not slide. Push the first glider as you did in the trial run. Observe the second glider and write down your observations. Record measured velocities in Table 10-1.

Table 10-1 Velocities in elastic collisions of equal masses.

Trial	Velocity of Glider 1 "Before" v_1	Velocity of Glider 2 "Before" v_2	Velocity of Glider 1 "After" u_1	Velocity of Glider 2 "After" u_2
1		0		
2		0		
3		0		
4		0		

Analysis of the Results

Did you notice any pattern in the distribution of velocities u_1 and u_2 after the gliders collide? Describe it here.

What causes the experimental error in this exercise? Can you eliminate it?

Part 2. Elastic Collisions, $m_1 > m_2$

In this part of the lab, the second glider will remain at rest, $v_2 = 0$. You will continue testing collisions between gliders of unequal masses, with the first glider mass being larger than the glider at rest (see Diagram 2).

Prediction Based on the Theory

The law of conservation of linear momentum adapted for this part of the lab exercise looks like the following:

$$m_1 v_1 = m_1 u_1 + m_2 u_2$$

According to the previous equation, the momentum carried by the first glider should be distributed somehow between glider 1 and glider 2 after the collision. Is it possible to predict the distribution of linear momentum based on the previous equation? Discuss with your group. Write down your conclusions.

Testing

Add mass to the first glider and record it in Table 10-2. Repeat the experimental procedure described in Part 1.

In the next few trials, observe the transfer of linear momentum when the gliders collide. Write down your observations.

Record the velocities before and after the collision in Table 10-2a. Calculate the momenta before and after the collisions in Table 10-2b.

Table 10-2a Velocities in elastic collisions of unequal masses.

Trial	Mass m_1	Velocity of Glider 1 "Before" v_1	Velocity of Glider 2 "Before" v_2	Velocity of Glider 1 "After" u_1	Velocity of Glider 2 "After" u_2
1			0		
2			0		
3			0		
4			0		

Table 10-2b Linear momenta in elastic collisions of unequal masses.

Trial	Linear Momentum of the System "Before" p_{before}	Linear Momentum of Glider 1 "After" p_1	Linear Momentum of Glider 2 "After" p_2	Linear Momentum of the System "After" p_{after}
1				
2				
3				
4				

Analysis of the Results

Notice the distribution of linear momentum after the collision. Write down your observations.

Part 3. Elastic Collisions, $m_1 < m_2$

In this part of the lab, the second glider will remain at rest, $v_2 = 0$. You will continue testing collisions between gliders of unequal masses, with the second glider mass being larger than the glider initially in motion (see Diagram 3).

Prediction Based on the Theory

The law of conservation of linear momentum adapted for this part of the lab exercise looks like the following:

$$m_1 v_1 = m_1 u_1 + m_2 u_2$$

According to the previous equation, the momentum carried by the first glider should be distributed somehow between glider 1 and glider 2 after the collision. Is it possible to predict the distribution of linear momentum based on the previous equation? Discuss with your group. Write down your conclusions.

Testing

Add mass to the second glider (the one at rest) and record it in Table 10-3a. Repeat the experimental procedure.

In the next few trials, observe the transfer of linear momentum when the gliders collide. Write down your observations.

Record the velocities before and after the collision in Table 10-3a. Calculate the momenta before and after the collisions in Table 10-3b.

Table 10-3a Velocities in elastic collisions of unequal masses.

Trial	Mass m_2	Velocity of Glider 1 "Before" v_1	Velocity of Glider 2 "Before" v_2	Velocity of Glider 1 "After" u_1	Velocity of Glider 2 "After" u_2
1			0		
2			0		
3			0		
4			0		

Table 10-3b Linear momenta in elastic collisions of unequal masses.

Trial	Linear Momentum of the System "Before" p_{before}	Linear Momentum of Glider 1 "After" p_1	Linear Momentum of Glider 2 "After" p_2	Linear Momentum of the System "After" p_{after}
1				
2				
3				
4				

Analysis of the Results

Notice the distribution of linear momentum after the collision. Write down your observations.

Part 4. Elastic Collisions, $m_1 = m_2$

In this part of the lab, you will test collisions between gliders of equal masses when both gliders have initial velocities (see Diagram 4).

Prediction Based on the Theory

The law of conservation of linear momentum adapted for this part of the lab exercise looks like the following:

$$m_1v_1 + m_2v_2 = m_1u_1 + m_2u_2$$

In this part of the lab, $m_1 = m_2$, so we can eliminate mass from the equations by dividing both sides by m:

$$v_1 + v_2 = u_1 + u_2$$

According to the previous equation, the velocities of both gliders before the collision should be distributed between glider 1 and glider 2 after the collision. Is it possible to predict the distribution of linear momentum based on the previous equation? Discuss with your group. Write down your conclusions.

Testing

In the next few trials, observe the transfer of linear momentum when the gliders collide. Write down your observations.

Record the velocities before and after the collision in Table 10-4a. Calculate the momenta before and after the collisions in Table 10-4b.

Table 10-4a Velocities in elastic collisions.

Trial	Velocity of Glider 1 "Before" v_1	Velocity of Glider 2 "Before" v_2	Velocity of Glider 1 "After" u_1	Velocity of Glider 2 "After" u_2
1				
2				
3				
4				

Table 10-4b Linear momenta in elastic collisions.

Trial	Linear Momentum of Glider 1 "Before" p_1	Linear Momentum of Glider 2 "Before" p_2	Linear Momentum of the System "Before" p_{before}
1			
2			
3			
4			

Trial	Linear Momentum of Glider 1 "After" p_1	Linear Momentum of Glider 2 "After" p_2	Linear Momentum of the System "After" p_{after}
1			
2			
3			
4			

Analysis of the Results

Notice the distribution of linear momentum after the collision. Write down your observations.

Part 5. Inelastic Collisions, $m_1 = m_2$

In this part of the lab, the second glider will remain at rest, $v_2 = 0$. Use the Velcro bumpers to model inelastic collisions (see Diagram 5).

Prediction Based on the Theory

The law of conservation of linear momentum adapted for this part of the lab exercise looks like the following:

$$m_1 v_1 = (m_1 + m_2) \cdot u$$

Notice that, after the collision, there is only one time measured (the gliders move together).

Is it possible to predict the value of velocity of the system and its linear momentum based on the previous equation; for example, by eliminating mass from the equation? Discuss with your group. Write down your conclusions.

Testing

In the next few trials, observe the transfer of linear momentum when the gliders collide. Write down your observations.

Record the velocities before and after the collision in Table 10-5. Calculate the momenta before and after the collisions.

Table 10-5 Velocities in elastic collisions of unequal masses.

Trial	Velocity of Glider 1 "Before" v_1	Velocity of Glider 2 "Before" v_2	Linear Momentum of the System "Before" p_{before}	Velocity of Both Gliders "After" u	Linear Momentum of the System "After" p_{after}
1		0			
2		0			
3		0			
4		0			

Analysis of the Results

Write down your observations.

Synthesis

Use arrows to mark the velocities/momenta of the gliders after collisions in the tested scenarios. Indicate the direction and magnitude of the velocities.

Part 1
Elastic collision
$m_1 = m_2$
$v_1 > 0, v_2 = 0$

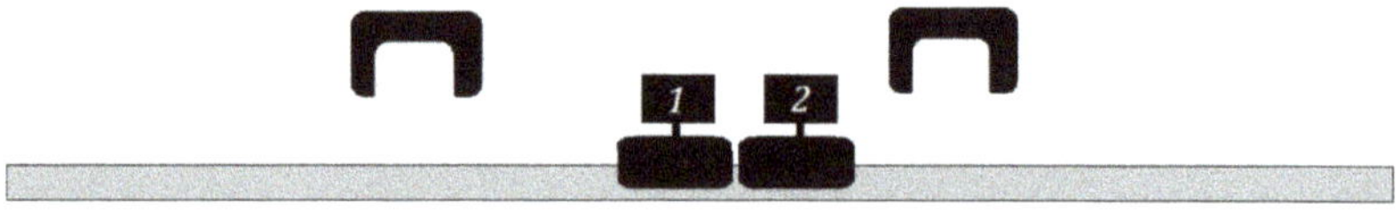

Part 2
Elastic collision
$m_1 > m_2$
$v_1 > 0, v_2 = 0$

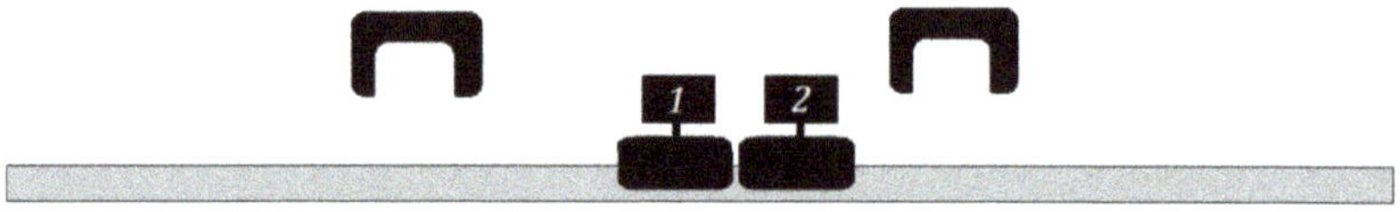

Part 3
Elastic collision
$m_1 < m_2$
$v_1 > 0, v_2 = 0$

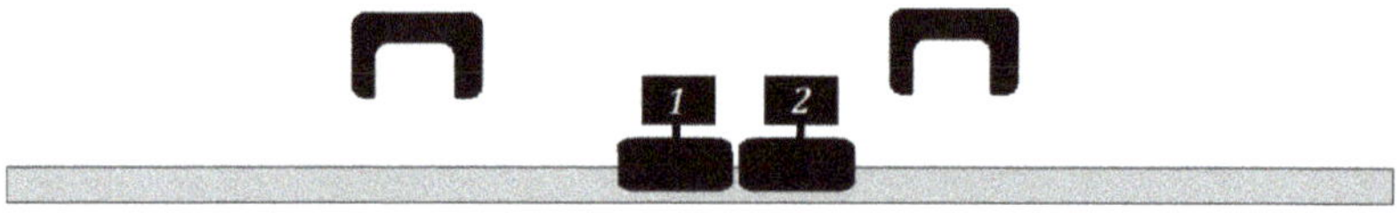

Part 4
Elastic collision
$m_1 = m_2$
$v_1 > 0, v_2 > 0$

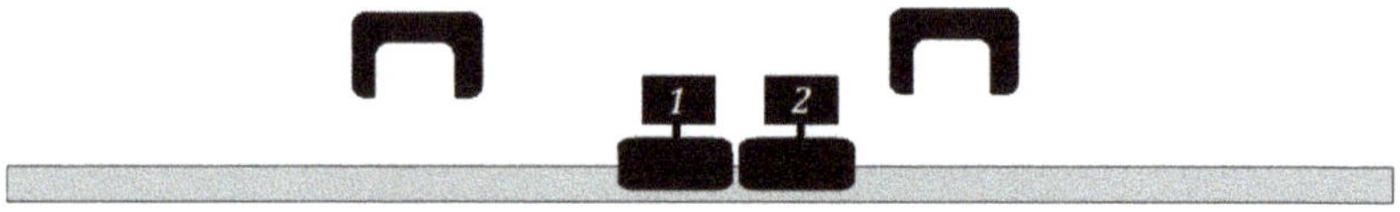

Part 5
Inelastic collision
$m_1 = m_2$
$v_1 > 0, v_2 = 0$

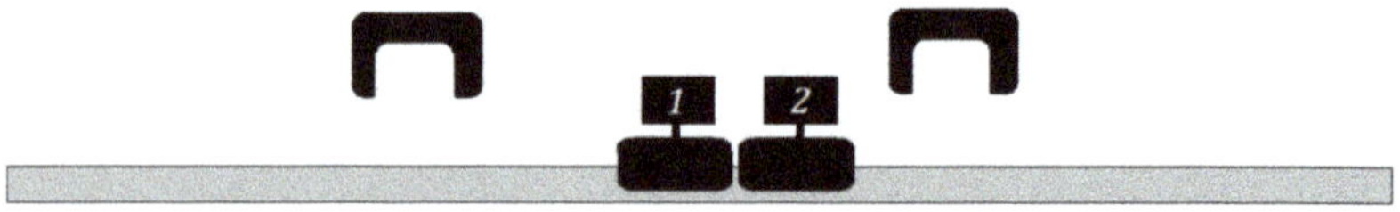

Reflection and Evaluation

This was a very long lab exercise. Congratulations, you successfully completed it!

Rate the following:

Your preparedness	☆☆☆☆☆	In-group communication	☆☆☆☆☆
Your engagement	☆☆☆☆☆	Group work	☆☆☆☆☆
Your time management	☆☆☆☆☆	Group effectiveness	☆☆☆☆☆

Lab 11

Torque and Rotations

Introduction

The purpose of this lab is to test rotational equilibrium.

Upon completion of this lab, students will

- Define torque.
- Describe the effect of the off-center forces.
- Find the center of mass.

Theory

1. What is torque?

2. What conditions must be met for an object to remain in rotational equilibrium?

3. What conditions must be met for an object to remain in translational equilibrium?

4. Define the center of mass.

5. Fill in the blanks.

Force applied centrally causes _________________ motion.

Force applied off-center causes _________________ motion.

To balance an object, support it at ______________________.

Torque is a ___________________ quantity.

center of mass	rotational	scalar	translational	vector

Experimental Settings

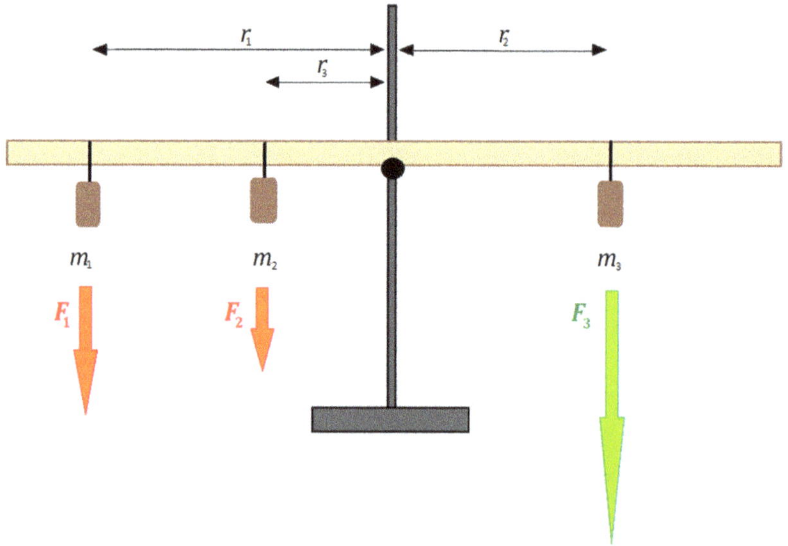

Figure 11-1 Experimental settings diagram.

1. Set the yardstick and the weights as shown in Figure 11-1.
2. Make sure that the weights are hanging securely above the lab table. Otherwise, somebody might get hurt if any of the weights drops down on the floor.
3. Perform a few trials to see how changing the weight distribution affects the equilibrium.

Part 1. Finding the Third Mass

In this part of the lab, masses m_1 and m_2 are given along with their arms. Find the third mass that would set the system in equilibrium.

Let r_1 = 40 units and m_1 = 100 g. Let r_2 = 20 units and m_2 = 100 g. What mass should be suspended at r_3 = 30 units to set the system in equilibrium?

m_3 = ____________________

Let r_1 = 30 units and m_1 = 110 g. Let r_2 = 20 units and m_2 = 110 g. What mass should be suspended at r_3 = 55 units to set the system in equilibrium?

m_3 = ____________________

Let r_1 = 40 units and m_1 = 100 g. Let r_2 = 20 units and m_2 = 150 g. Where should m_3 = 250 g be suspended to set the system in equilibrium?

m_3 = ____________________

Let r_1 = 30 units and m_1 = 100 g. Let r_2 = 20 units and m_2 = 150 g. Where should m_3 = 250 g be suspended to set the system in equilibrium?

m_3 = ____________________

Part 2. Balancing Three Masses

In this part of the lab, only the masses are given. Find the arms to set the system in rotational equilibrium.

Let m_1 = 100 g, m_2 = 100 g, and m_3 = 100 g. Where should the masses be suspended to set the system in rotational equilibrium?

r_1 = ________________ r_2 = ________________ r_3 = ________________

Let m_1 = 100 g, m_2 = 200 g, m_3 = 150 g. Where should the masses be suspended to set the system in rotational equilibrium?

r_1 = ________________ r_2 = ________________ r_3 = ________________

Let m_1 = 150 g, m_2 = 250 g, and m_3 = 100 g. Where should the masses be suspended to set the system in rotational equilibrium?

r_1 = ________________ r_2 = ________________ r_3 = ________________

Quick Evaluation

	Strongly agree	Agree	Neutral	Disagree	Strongly disagree
It is easy to set the equilibrium.	o	o	o	o	o
I am doing my best.	o	o	o	o	o
I am learning something.	o	o	o	o	o
I contribute to the group work.	o	o	o	o	o
I have a feeling of accomplishment.	o	o	o	o	o

Part 3. Balancing the Yardstick

Support the yardstick off-center as shown in Figure 11-2.

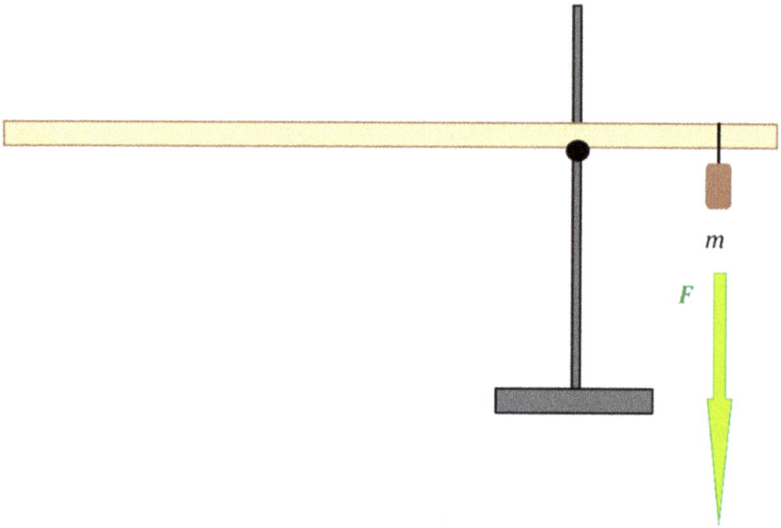

Figure 11-2 Yardstick supported off-center.

Find mass that equilibrates the longer part of the yardstick.

Propose an equation that describes the torques presented in Figure 11-2.

Part 4. The Art of Balancing

Design an interesting torque balance on your own. Write an equation that represents your designed balance. You can use the lab set provided or any objects of daily use you can find around. Send photographic documentation of it to the instructor.

Part 5. Center of Mass

Find the center of mass of the provided object. Describe the method.

Draw the object in the space that follows. Mark the center of mass.

Lab 12
Buoyant Force

Introduction

The purpose of this lab is to investigate the buoyant force.

Upon completion of this lab, students will

- Know Archimedes' principle.
- Use experimental design to find the buoyant force.
- Explain how the buoyant force depends on the density of a liquid.

Theory

1. State Archimedes' principle.

2. Derive the formula to calculate buoyant force.

3. Explain specific gravity.

4. Describe a hydrometer and how it works

One-Minute Mindfulness Exercise

There is a legend about Archimedes who, after discovering the principle named after him, raced through his town crying, "Eureka! Eureka!"

Have you ever experienced such a moment of enlightenment? Share with your lab team.

Part 1. Buoyant Force

In this part of the lab, you will measure the buoyant force of an object in various liquids.

Step One. Find the volume of the object. Describe the method of measuring the volume in the space that follows.

V_{object} = ______________________

Step Two. Use the force sensor or a hanging scale to weigh the object in air (Figure 12-1a).

$W_{object\ in\ air}$ = ______________________

a. b.

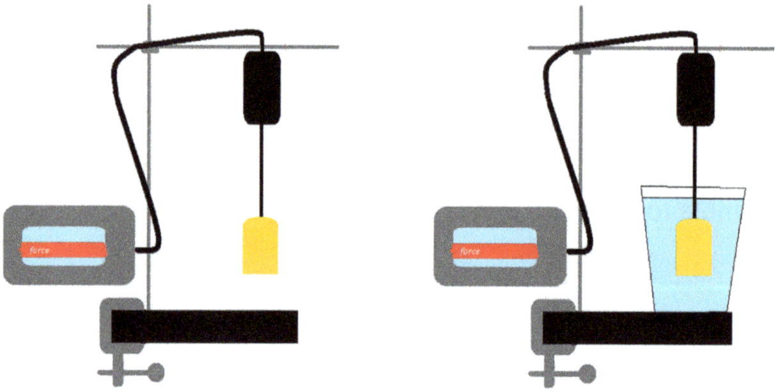

Figure 12-1 Experimental settings.

Step Three. Weigh the object completely submerged in a liquid, as shown in Figure 12-1b. Collect the measurements in Table 12-1.

Table 12-1 Data collection.

Liquid	Density of the Liquid	Weight of the Submerged Object	Buoyant Force	Weight of the Displaced Liquid	Percent Difference
Column 1	Column 2	Column 3	Column 4	Column 5	Column 6
water	1 g/cm^3				
oil					

Step Four. Calculate the buoyant force B according to the following formula. Record the results in Column 4 of Table 12-1.

$$B = W_{object\ in\ air} - W_{object\ in\ liquid}$$

Step Five. Use the volume of the object and density of the liquid to calculate the weight of the displaced liquid. Record the results in Column 5 of Table 12-1.

$$W_{displaced\ liquid} = m_{displaced\ liquid} \cdot g = V_{object} \cdot \rho_{liquid} \cdot g$$

Step Six. Compare the reading on the measured buoyant force (Column 4) and your calculations (Column 5) by finding the percent difference. Record the results in Column 6 of Table 12-1.

$$\frac{\left|B - W_{displaced\ liquid}\right|}{(B + W)/2} \cdot 100\%$$

- If the difference is greater than 5%, consider repeating the experimental procedure.
- If the difference is less than or equal to 5%, list the sources of error.

Part 2. Floating Block

In this part of the lab, you will

A. Find the density of the wood floating in water.
B. Experimentally balance the buoyant force.

Experimental Settings A

Place a wooden block in water and let it float as shown in Figure 12-2.

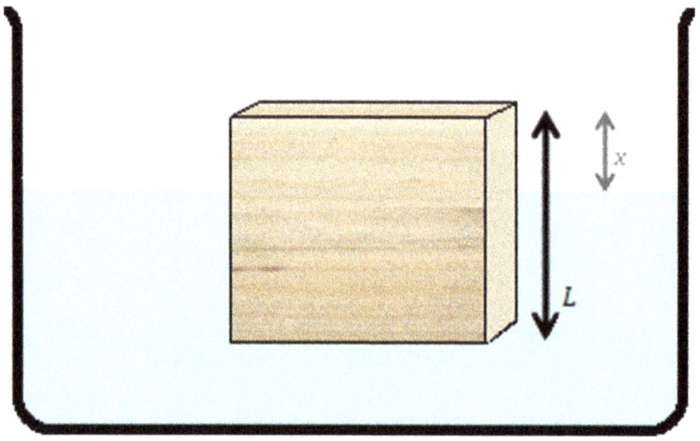

Figure 12-2 Wooden block floating in water.

If the block floats, $F_{net} = 0$

$$W_{wooden\ block} = B$$

$$m_{wooden\ block} \cdot g = m_{displaced\ water} \cdot g$$

$$m_{wooden\ block} = m_{displaced\ water}$$

$$\rho_{wood} \cdot V_{wooden\ block} = \rho_{water} \cdot V_{displaced\ water}$$

$$\rho_{wood} \cdot (L \cdot A) = \rho_{water} \cdot ([L - x] \cdot A)$$

$$\rho_{wood} \cdot L = \rho_{water} \cdot (L - x)$$

$$\rho_{wood} = \rho_{water} \cdot \frac{L - x}{L}$$

What measurements are needed?

How will you take the measurements?

ρ = ______________________

Experimental Settings B

Place some mass on the wooden block floating in water as shown in Figure 12-3.

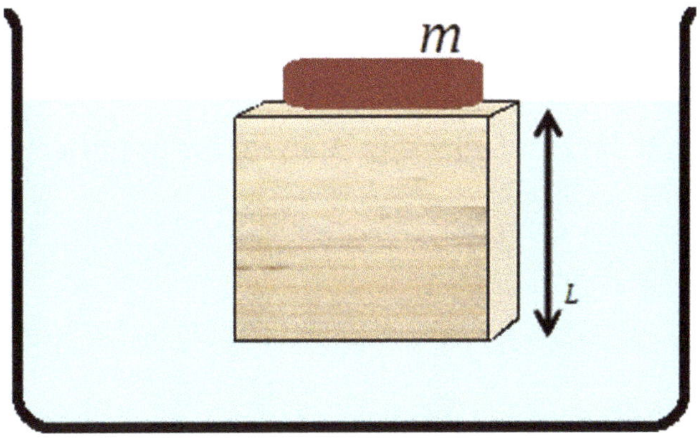

Figure 12-3 Wooden block floating in water.

Step One. Use the trial and error method to find what mass should be placed to submerge the wooden block in water but not to sink it.

m_x = ___________________

Step Two. Calculate the mass and compare it with the experimental results.

If the block floats, $F_{net} = 0$

$$W_{wooden\ block} + W_x = B$$

$$m_{wooden\ block} \cdot g + m_x \cdot g = m_{displaced\ water} \cdot g$$

$$m_{wooden\ block} + m_x = m_{displaced\ water}$$

$$\rho_{wood} \cdot V_{wooden\ block} + m_x = \rho_{water} \cdot V_{displaced\ water}$$

Since the entire block is submerged under water, the volume of the displaced water equals the volume of the block.

$$m_x = \rho_{water} \cdot V - \rho_{wood} \cdot V$$

$$m_x = \left(\rho_{water} - \rho_{wood}\right) \cdot V$$

m_x = ___________________

Part 3. Buoyant Force as a Function of Density of Liquid

Design an experiment to investigate how the density of the liquid changes the buoyant force.

- Describe the step-by-step experimental procedure.
- Find at least eight data points.
- Plot the results. Find the best fit line and interpret the coefficients.

Lab 13
Springs

Introduction

The purpose of this lab is to investigate linear and nonlinear elastic materials, to find the spring constant for two different springs, and to investigate the spring constant of systems of springs connected in series and parallel.

Upon completion of this lab, students will

- Know the difference between linear and nonlinear elastic materials.
- Recognize the Hookean materials based on the force-elongation graph.
- Understand Hooke's law.
- Explain the behavior of linear and nonlinear elastic materials based on their molecular structure.
- Derive mathematical formulas based on experimental data.

Theory

Hooke's law states that the restoring force of the spring is proportional to its elongation. Using mathematics, it can be written as

$$F = -k \cdot x$$

where k is a spring constant that describes physical properties of the spring.

Notice the minus in the force formula. It simply indicates that the restoring force is always opposite to the stretching force. According to Newton's third law, the magnitude of the restoring force is the same as the magnitude of the external force applied to stretch the spring.

Materials that obey the law are called Hookean or linear elastic.

Experimental Settings

Set a spring at the lab stand as shown in Figure 13-1. To minimize vibrations, use a C-clamp to attach the stand to the table.

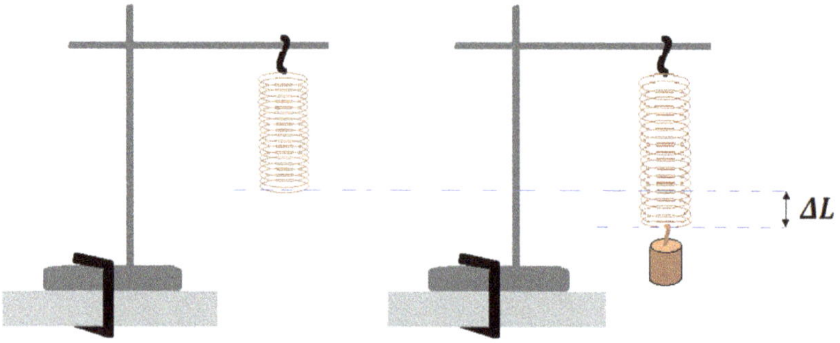

Figure 13-1 Hooke's law experimental settings.

In this lab, you will measure the elongation of the spring (not its length), labeled as ΔL in Figure 13-1.

Discuss with your group how to measure the elongation. Hint: Decide the point of reference from which you will measure the change of position of the lower end of the spring. You can choose the horizontal bar of the lab stand, the table surface, or the floor. Plan how to minimize the error.

In Figure 13-1, sketch how you plan to measure the elongation of the spring.

Part 1. Linear Elastic Materials

Here is the experimental procedure step-by-step:

Step One. Add the smallest mass to the spring. Record the mass in Table 13-1, Column 1. Calculate the applied force.

$$F = m \cdot g$$

and record it in Column 2.

Step Two. Measure the elongation. Record the measurement in Column 3.

Table 13-1 Elongation of the spring.

Mass [kg]	Force [N]	Spring 1 ΔL [m]	Spring 2 ΔL [m]
Column 1	Column 2	Column 3	Column 4

Step Three. Repeat the measurements for another spring and record it in Column 4.

Data Analysis

Plot the data on Graph 13-1. Compare the patterns. Write down your observations.

Perform regression to find the best fit line for both sets of points. Record spring constant k for both springs.

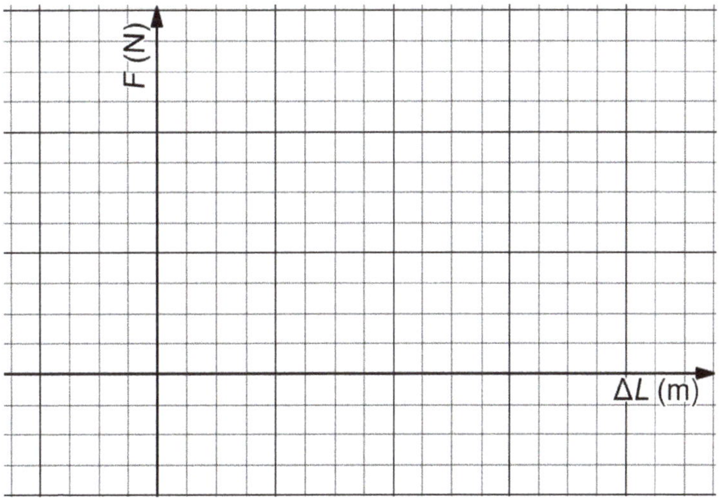

Graph 13-1 Linear elastic materials.

k_1 = ____________________

k_2 = ____________________

Part 2. Nonlinear Elastic Materials

Step One. Replace the spring with a rubber band and repeat the measurements.

Step Two. Begin with weight that stretches the rubber band. Record the mass and calculate the applied force.

$$F = m \cdot g$$

Step Three. Repeat the measurements for another rubber band and record the data in Table 13-2.

Table 13-2 Elongation of the rubber band.

Mass [kg]	Force [N]	Rubber Band 1 ΔL [m]	Rubber Band 2 ΔL [m]

Data Analysis

Plot the data on Graph 13-2. Compare the patterns. Write down your observations.

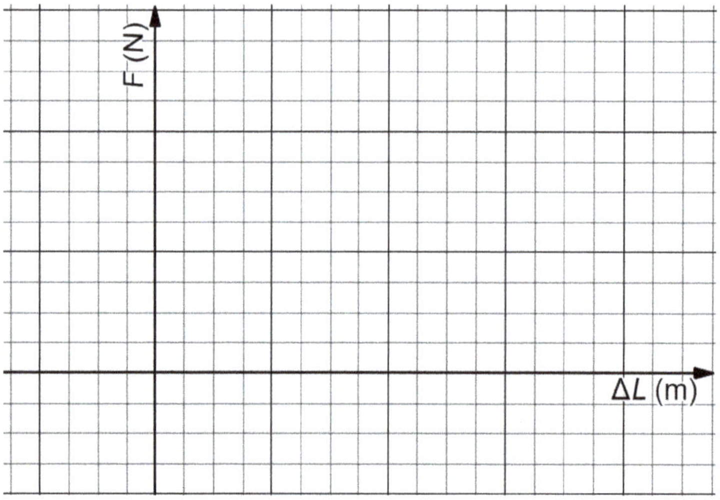

Graph 13-2 Nonlinear elastic materials.

1. List similarities and differences between Graphs 13-1 and 13-2.

2. What function is the best to approximate the relation between the stretch and the force based on the plotted points?

Part 3. Combining Springs

In this part of the lab, you will combine two identical springs first in series (Figure 13-2a) and then in parallel (Figure 13-2b).

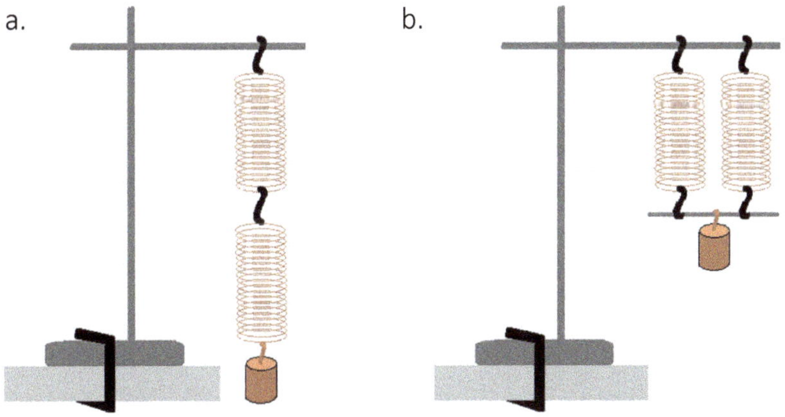

Figure 13-2 Springs connected in series (a) and parallel (b).

Think about the meaning of the spring constant and try to predict the following:

1. What is a spring constant for a set of similar springs connected in series? Is it the same as it would be for one spring? Is it double as the single spring constant? Is it halved? Why?

2. What is a spring constant for a set of similar springs connected in series? Is it the same as for one spring? Is it doubled? Is it halved? Why?

Check your predictions experimentally and record the results in Table 13-3.

Table 13-3 Elongation of the spring.

Mass [kg]	Force [N]	Springs in Series ΔL [m]	Springs in Parallel ΔL [m]

Data Analysis

Plot the data on Graph 13-3. Compare the patterns. Write down your observations.

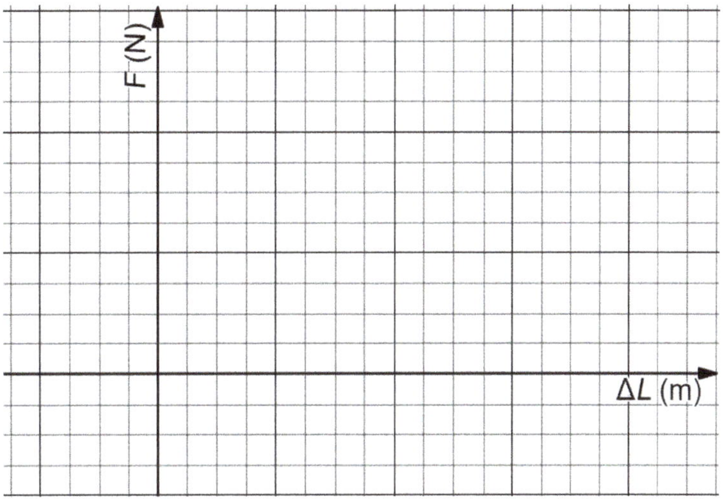

Graph 13-3 Springs in series and parallel.

Perform regression to find the best fit line for both sets of points.

Results:

k_{series} = ________________________

$k_{parallel}$ = ________________________

Discussion

- Were your predictions about the spring constant of the combinations of springs correct?
- Was there anything surprising in this part of your lab work?
- What would be the spring constant of three springs combined in series? What if they were combined in parallel?

Discuss in your group.

Lab 14
Simple Harmonic Motion

Introduction

The purpose of this lab is to investigate simple harmonic motion of a spring-mass system.

Upon completion of this lab, students will

- Know the characteristics of periodic motion.
- Model the period of the simple harmonic oscillator with a trigonometric function.
- Interpret the parameters of sine/cosine function describing harmonic oscillations.
- Analyze the period of the spring-mass system as a function of mass.

Experimental Settings

In this lab, we will continue working with springs. Recall the spring constants from Lab 13.

Spring 1 k_1 = ____________________

Spring 2 k_2 = ____________________

Set the spring on the hook of the force sensor. Secure the stand with a C-clamp and attach the smallest mass, as shown in Figure 14-1.

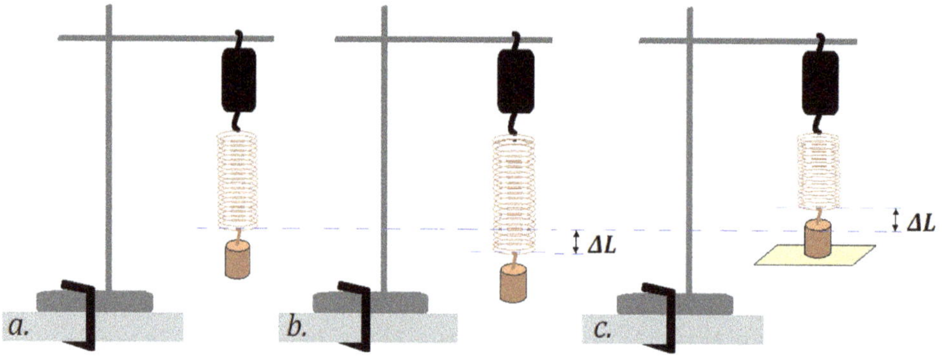

Figure 14-1 Spring-mass system experimental settings: (a) spring-mass system in equilibrium, (b) elongated spring, (c) relaxed spring.

Examination of the Spring-Mass System

1. When the system is in equilibrium, the spring is stretched just enough to uphold the suspended weight (Figure 14-1a). Write down the equation of forces in equilibrium.

2. Pulling down the weight stretches the spring (Figure 14-1b). Write down the equation of forces.

3. Supporting weight relaxes the spring (Figure 14-1c). Write down the equation of forces.

4. Gently stretch the spring down and see it bouncing up. Observe the motion of the weight. Is the elongation of the spring the same as the height it bounces up in its harmonic oscillation? Why?

5. Support the weight so the spring is in a relaxed position. Release the weight. Which graph in Figure 14-2 represents the position of the weight attached to the spring if you begin your observation when the spring is in the relaxed position?

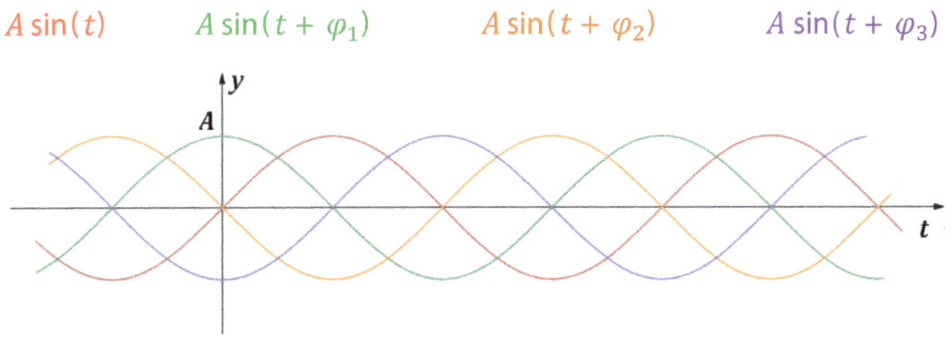

Figure 14-2 Four sin(x) functions shifted in phase φ.

6. Based on your observations, mark points of the highest velocity on the graph of position as a function of time for a mass-spring system in Figure 14-3.

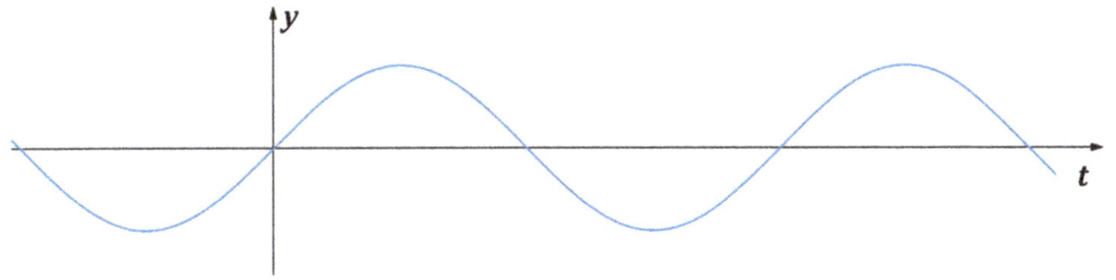

Figure 14-3 Graph of a position of mass versus time in simple harmonic oscillations.

7. Mark the period of oscillations in Figure 14-3.

Theory

The period of oscillation of a spring-mass oscillator is given by the following formula:

$$T = 2\pi\sqrt{\frac{m}{k}}$$

where k is the spring constant (see Lab 13) and m is the mass suspended on the spring.

Based on the formula, sketch the graph of a mathematical function that represents the period of a harmonic oscillator as a function of suspended mass, $T = T(m)$, in the Cartesian plane in Graph 14-1.

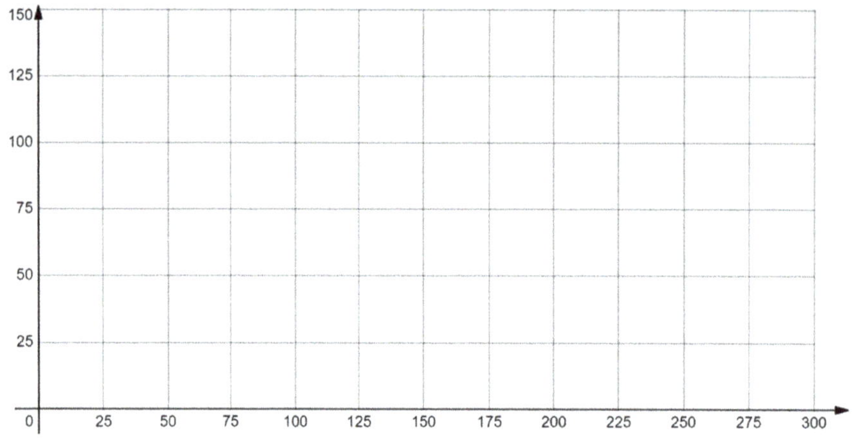

Graph 14-1 The period of a harmonic oscillator as a function of suspended mass.

Compare your graph to those proposed by other students in your group.

Synthesis

The period of a simple pendulum is given by the following formula:

$$T = 2\pi\sqrt{\frac{l}{g}}$$

and the period of a harmonic oscillator is

$$T = 2\pi\sqrt{\frac{m}{k}}$$

Discuss similarities and differences between those formulas in your group. The ideas for discussion might include, but are not limited to:

- Mathematical operations in both formulas
- Numbers
- Units
- Graphs of motion
- Physical properties of both objects/phenomena described by the formulas
- Mathematical functions used to model the motion of both

As a group, fill out the following diagram.

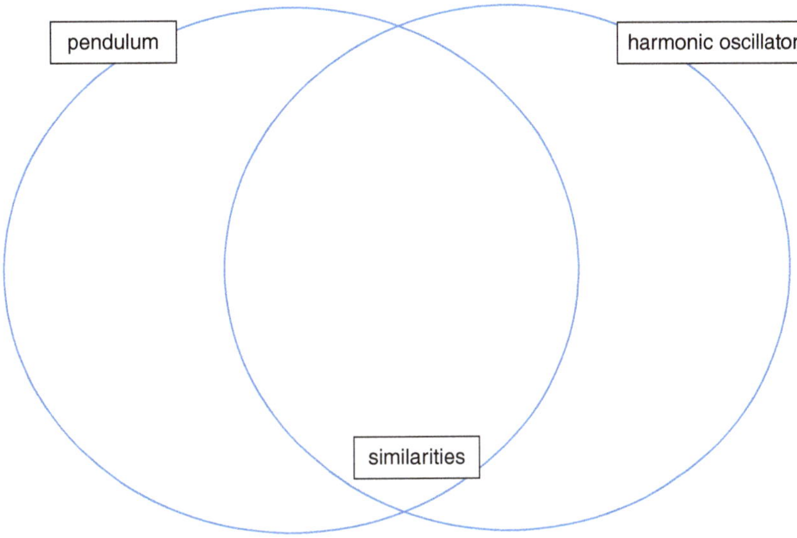

Part 1. Analysis of the Period of Oscillation as a Function of Mass

Step One. Choose a spring and record the spring constant.

k_1 = ____________________

With the smallest mass suspended on the spring, gently pull the weight down to move the system from its equilibrium position.

Step Two. When the oscillations stabilize (you will observe no swings or twists, just smooth oscillations about the position of equilibrium), measure time for 10 oscillations.

Step Three. Collect data in Table 14-1. Record the mass suspended on the spring (Column 1) and the time of 10 oscillations (Column 2). Divide the measured time by the number of oscillations (10) to obtain the period of oscillation and record it in Column 3.

Step Four. Calculate the period of oscillation based on the previous formula and record it in Column 4.

Step Five. Calculate the percentage difference between the measured and calculated values and record it in Column 5.

$$\frac{|T_c - T_m|}{T_c} \cdot 100\%$$

Repeat Steps Two to Five with increasing masses.

Table 14-1 Period of oscillation.

	m [kg]	t [s]	$T_{measured}$ [s]	$T_{calculated}$ [s]	Percent Difference
Trial	Column 1	Column 2	Column 3	Column 4	Column 5
1					
2					
3					
4					
5					
6					
7					
8					

Step Six. Choose a spring and record the spring constant.

$$k_2 = \_\_\_\_\_\_\_\_\_\_\_\_\_\_\_\_$$

With the smallest mass suspended on the spring, gently pull the weight down to move the system from its equilibrium position. Repeat Steps Two to Five with increasing masses and record the data in Table 14-2.

Table 14-2 Period of oscillation.

	m [kg]	*t* [s]	$T_{measured}$ [s]	$T_{calculated}$ [s]	Percent Difference
Trial	Column 1	Column 2	Column 3	Column 4	Column 5
1					
2					
3					
4					
5					
6					
7					
8					

Step Seven. Plot the data collected in Table 14-1 and Table 14-2 into one graph below.

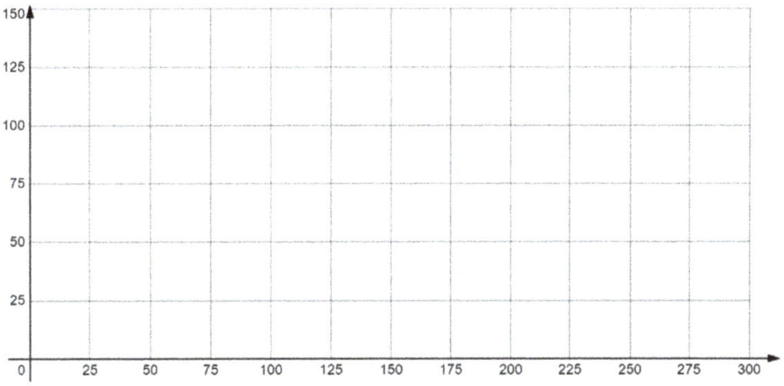

Compare both graphs. Are they the same? What causes the differences, if any?

Compare your graph with the one sketched in the Theory part. Write down your observations.

Force Equation

Use the "Analyze" function of the ***LabQuest*** interface to find an equation that describes changes in the force of the spring. The mathematical function that describes the force of the spring is presented as

$$y = A \cdot sin\left(Bx + C\right) + D$$

Write down the respective coefficients A, B, C, and D in Table 14-3.

Table 14-3 Force equation coefficients, spring 1.

	Mass [kg]	*A*	*B*	*C*	*D*
Trial	Column 1	Column 2	Column 3	Column 4	Column 5
1					
2					
3					
4					
5					
6					
7					
8					

Discussion

1. Are there any patterns that indicate relationships between columns, rows, or cells in Table 14-3? Do you notice increasing, decreasing, or constant values?

2. Is there any relationship between coefficients A, B, C, and D and suspended mass?

Table 14-4 Force equation coefficients, spring 2.

	Mass [kg]	*A*	*B*	*C*	*D*
Trial	**Column 1**	**Column 2**	**Column 3**	**Column 4**	**Column 5**
1					
2					
3					
4					
5					
6					
7					
8					

Discussion

1. Are there any patterns that indicate relationships between columns, rows, or cells in Table 14-4? Do you notice increasing, decreasing, or constant values?

2. Is there any relationship between coefficients *A*, *B*, *C*, and *D* and suspended mass? Is it like the one observed with spring 1?

3. How do you interpret the coefficients?

A denotes ______________________________

B stands for ______________________________

C represents ______________________________

D can be interpreted as ______________________________

Evaluation of the Results

Compare the data collected in both methods in Table 14-5. Follow the step-by-step instructions that follow:

Step One. Column 1. Copy the values from Table 14-1, Column 3.

Step Two. Column 2. Calculate the period of oscillation based on the B coefficient values collected in Table 14-3, Column 3.

$$T = \frac{2\pi}{B}$$

Step Three. Column 3. Find the percentage difference between T_B and T_k.

Table 14-5 Comparison of data obtained with two methods (spring 1).

	T_k	T_B	Percent Difference
Trial	Column 1	Column 2	Column 3
1			
2			
3			
4			
5			
6			
7			
8			

If the percent difference (Column 3) is greater than 3% but less than or equal to 5%, identify the sources of error. If the percent difference is greater than 5%, identify the sources of error and repeat the measurement.

Repeat the comparison for data obtained for spring 2 and record it in Table 14-6.

Table 14-6 Comparison of data obtained with two methods (spring 2).

	T_k	T_B	**Percent Difference**
Trial	**Column 1**	**Column 2**	**Column 3**
1			
2			
3			
4			
5			
6			
7			
8			

If the percent difference is greater than 3% but less than or equal to 5%, identify the sources of error.

If the percent difference is greater than 5%, identify the sources of error and repeat the measurement.

One-Minute Mindfulness Exercise

This was a long and monotonous lab. Sometimes science is like that. You work for years collecting datasets, observing them, learning their behavior, and recognizing the patterns. I do not think there is a shortcut. Natural phenomena and the laws that govern them remain on the textbook level before you get tired in the lab.

www.ingramcontent.com/pod-product-compliance
Ingram Content Group UK Ltd.
Pitfield, Milton Keynes, MK11 3LW, UK
UKHW051306160926
15216UKWH00010B/24

9 798385 100903